ENERGY POLICIES, POLITICS AND PRICES

TRANSPORTING CRUDE OIL BY U.S. RAIL

BACKGROUND, ISSUES, AND TRENDS

ENERGY POLICIES, POLITICS AND PRICES

Additional books in this series can be found on Nova's website under the Series tab.

Additional e-books in this series can be found on Nova's website under the e-book tab.

ENERGY POLICIES, POLITICS AND PRICES

TRANSPORTING CRUDE OIL BY U.S. RAIL

BACKGROUND, ISSUES, AND TRENDS

ROSARIO S. MCLAUGHLIN
EDITOR

nova publishers
New York

For permission to use material from this book please contact us:
Telephone 631-231-7269; Fax 631-231-8175
Web Site: http://www.novapublishers.com

Additional color graphics may be available in the e-book version of this book.

Library of Congress Cataloging-in-Publication Data

ISBN: 978-1-63117-837-5

Published by Nova Science Publishers, Inc. † New York

CONTENTS

PREFACE

North America is experiencing a boom in crude oil supply, primarily due to growing production in the Canadian oil sands and the recent expansion of shale oil production from the Bakken fields in North Dakota and Montana as well as the Eagle Ford and Permian Basins in Texas. Taken together, these new supplies are fundamentally changing the U.S. oil supply-demand balance. The United States now meets 66% of its crude oil demand from production in North America, displacing imports from overseas and positioning the United States to have excess oil and refined products supplies in some regions. This book provides a background of the United States rail transportation of crude oil, as well as discusses the issues it leaves for Congress.

Chapter 1 – North America is experiencing a boom in crude oil supply, primarily due to growing production in the Canadian oil sands and the recent expansion of shale oil production from the Bakken fields in North Dakota and Montana as well as the Eagle Ford and Permian Basins in Texas. Taken together, these new supplies are fundamentally changing the U.S. oil supply-demand balance. The United States now meets 66% of its crude oil demand from production in North America, displacing imports from overseas and positioning the United States to have excess oil and refined products supplies in some regions. The rapid expansion of North American oil production has led to significant challenges in transporting crudes efficiently and safely to domestic markets—principally refineries—using the nation's legacy pipeline infrastructure. In the face of continued uncertainty about the prospects for additional pipeline capacity, and as a quicker, more flexible alternative to new pipeline projects, North American crude oil producers are increasingly turning to rail as a means of transporting crude supplies to U.S. markets. According to rail industry officials, U.S. freight railroads are estimated to have carried more than 400,000 carloads of crude oil in 2013 (roughly equivalent to 280 million barrels), compared to 9,500 carloads in 2008. Crude imports by rail from Canada have increased more than 20-fold since 2011. While oil by rail has demonstrated benefits with respect to the efficient movement of oil from producing regions to market hubs, it has also raised significant concerns about transportation safety and potential impacts to the environment. The most recent data available indicate that railroads consistently spill less crude oil per ton-mile transported than other modes of land transportation. Nonetheless, safety and environmental concerns have been underscored by a series of major accidents across North America involving crude oil transportation by rail—including a catastrophic fire that caused numerous fatalities and destroyed much of Lac Mégantic, Quebec, in 2013. Following that event, the U.S. Department of Transportation issued a safety alert warning that the type of crude oil being transported from the Bakken region may be more flammable than traditional heavy crude oil. Legislation introduced in Congress following the Lac Mégantic disaster would require railroads to have at least two crew members aboard all trains. In addition, policymakers are discussing regulatory changes involving tank car design, prevention of derailments, and selection of preferred routes for transporting oil by rail. Congress may evaluate these changes in the reauthorization of the Rail Safety Improvement Act of 2008 (P.L. 110-432).

Chapter 2 – This is statement of Joseph C. Szabo, Administrator, Federal Railroad Administration..

Chapter 3 – This is statement of Cynthia L. Quarterman, Administrator, Pipeline and Hazardous Materials Safety Administration.

Chapter 4 – This is statement of Christopher A. Hart, Vice Chairman, National Transportation Safety Board.

Chapter 5 – This is statement of Geoffrey C. Blackwell, Chief, Office of Native Affairs and Policy, Federal Communications Commission.

Chapter 6 – Testimony of Edward R. Hamberger, President & CEO, Association of American Railroads.

Chapter 7 – Testimony of Bob Greco, Group Director of Downstream Operations, American Petroleum Institute.

Chapter 8 – Report of Emergency Order issued by the U.S. Department of Transportation on Emergency Restriction/Prohibition Order, Docket No. DOT-OST-2014-0025 [Standards to Transport Crude Oil by Rail], dated February 25, 2014.

In: Transporting Crude Oil by U.S. Rail
Editor: Rosario S. McLaughlin
ISBN: 978-1-63117-837-5

Chapter 1

U.S. RAIL TRANSPORTATION OF CRUDE OIL: BACKGROUND AND ISSUES FOR CONGRESS*

John Frittelli, Paul W. Parfomak, Jonathan L. Ramseur, Anthony Andrews, Robert Pirog and Michael Ratner

SUMMARY

North America is experiencing a boom in crude oil supply, primarily due to growing production in the Canadian oil sands and the recent expansion of shale oil production from the Bakken fields in North Dakota and Montana as well as the Eagle Ford and Permian Basins in Texas. Taken together, these new supplies are fundamentally changing the U.S. oil supply-demand balance. The United States now meets 66% of its crude oil demand from production in North America, displacing imports from overseas and positioning the United States to have excess oil and refined products supplies in some regions.

The rapid expansion of North American oil production has led to significant challenges in transporting crudes efficiently and safely to domestic markets—principally refineries—using the nation's legacy pipeline infrastructure. In the face of continued uncertainty about the prospects for additional pipeline capacity, and as a quicker, more flexible alternative to new

---

* This is an edited, reformatted and augmented version of a Congressional Research Service publication, CRS Report for Congress, R43390, dated February 6, 2014.

pipeline projects, North American crude oil producers are increasingly turning to rail as a means of transporting crude supplies to U.S. markets. According to rail industry officials, U.S. freight railroads are estimated to have carried more than 400,000 carloads of crude oil in 2013 (roughly equivalent to 280 million barrels), compared to 9,500 carloads in 2008. Crude imports by rail from Canada have increased more than 20-fold since 2011.

While oil by rail has demonstrated benefits with respect to the efficient movement of oil from producing regions to market hubs, it has also raised significant concerns about transportation safety and potential impacts to the environment. The most recent data available indicate that railroads consistently spill less crude oil per ton-mile transported than other modes of land transportation. Nonetheless, safety and environmental concerns have been underscored by a series of major accidents across North America involving crude oil transportation by rail—including a catastrophic fire that caused numerous fatalities and destroyed much of Lac Mégantic, Quebec, in 2013. Following that event, the U.S. Department of Transportation issued a safety alert warning that the type of crude oil being transported from the Bakken region may be more flammable than traditional heavy crude oil.

Legislation introduced in Congress following the Lac Mégantic disaster would require railroads to have at least two crew members aboard all trains. In addition, policymakers are discussing regulatory changes involving tank car design, prevention of derailments, and selection of preferred routes for transporting oil by rail. Congress may evaluate these changes in the reauthorization of the Rail Safety Improvement Act of 2008 (P.L. 110-432).

INTRODUCTION

North America is experiencing a boom in crude oil supply, primarily due to the growth of heavy crude production in the Canadian oil sands[1] and the recent expansion of shale oil production in North Dakota, Montana, and Texas. North American production now supplies 66% of U.S. crude oil demand, displacing crude from Latin America, Africa, and the Middle East.

This shift has led to significant challenges in transportation, as refineries that once received crude oil principally from oceangoing tankers are now seeing increasing deliveries by domestic transport. Existing pipeline capacity is, in some cases, insufficient to carry growing crude oil from some production areas, or does not link to the refineries needing the oil. The domestic barge network does not serve some key production regions located far from navigable waterways. As a quicker, more flexible alternative to new pipeline projects, North American crude oil producers are increasingly turning to rail as a means of transporting crude supplies to U.S. markets. Increased exports of refined products—and, if Congress changes the law, of crude oil—could lead to even larger volumes of oil being transported by rail. According to rail industry officials, U.S. freight railroads are estimated to have carried more than 400,000 carloads of crude oil in 2013, or roughly 280 million barrels, compared to 9,500 carloads in 2008.[2] Crude imports by rail from Canada have increased more than 20-fold since 2011.

The rapid increase in crude oil shipments by rail will likely increase the number of oil spills from rail transportation. However, the most recent data available indicate that railroads consistently spill less crude oil per ton-mile transported than other modes of land transportation. The amount of crude spilled per ton-mile of rail transport declined significantly between the early 1990s and the 2002-2007 period, the most recent years for which data are available.[3]

Nonetheless, the increase in rail shipments of crude has raised safety and environmental concerns. These concerns have been underscored by a series of major incidents involving crude oil transportation by rail, including a catastrophic fire and explosion in Lac Mégantic, Quebec, in July 2013 and a derailment in Casselton, ND, in December 2013 that led to a mass evacuation. Consequently, government agencies in the United States and Canada are considering new regulations related to oil transport by rail, and some Members of Congress have called for tighter rules governing crude oil railcars as well as a broader reconsideration of the role of rail in the nation's oil transportation infrastructure.[4]

WHY IS OIL MOVING BY RAIL?

In 2012, the United States produced 2.38 billion barrels of crude oil and imported another 3.10 billion barrels.[5] Canada has become the United States' leading foreign supplier, thanks to its increasing production from oil sands.[6] However, U.S. oil output has been increasing rapidly. In October 2013, U.S crude oil production exceeded imports for the first time since February 1995.[7]

The location of U.S. crude oil production has been changing rapidly. In particular, production in Alaska and from offshore sites has been declining, while production in Texas and North Dakota has been rising. The U.S. Geological Survey recently estimated that 2.7 billion barrels of light sweet crude oil remain in overlooked producing formations,[8] including the Eagle Ford shale, a prolific source of very light sweet crude oil in Texas, and the Bakken formation in North Dakota, a source of light sweet crude oil that rivals West Texas crude in quality.[9]

Almost all oil produced domestically, as well as some Canadian production, flows to one of the 115 U.S. refineries (**Figure 1**).[10] Nearly 45% of the country's refining capacity is located in the Gulf Coast, where 43 refineries process more than 9 million barrels of oil per day (bpd). However, the Midwest and the West Coast also have significant refining capacity.

The last entirely new petroleum refinery in the United States opened in 1976. The number of refineries in operation has steadily declined since then as refining capacity has become concentrated in ever larger refineries. A quarter of U.S. capacity is concentrated in 11 refineries with capacities exceeding 300,000 bpd. The largest, Shell/Motiva's Baytown, TX, refinery, was recently expanded to 600,000 bpd. Operable U.S. refining capacity has actually increased from 16.5 million to nearly 18 million bpd over the last decade. Refineries representing approximately 75% of domestic capacity (13.3 million bpd) have the ability to process heavy crude oils, but many smaller refineries can process only light to intermediate crude oil.

Each refinery depends upon a certain grade or blend of crude oils to operate efficiently, depending upon its custom-designed processing equipment. A refinery designed to run light crude oil could not switch to heavy crude oil without adding a coking unit, for example. However some refineries that process heavy sour crude could switch to lighter sweet crude by bypassing their coking units, if the economics of doing so are favorable. Until quite recently, the supply of light sweet crude oil was diminishing, but newly available light sweet crudes from North Dakota's Bakken formation are changing refining dynamics in some regions of the United States, especially as refineries seek supplies that cannot be delivered economically by tanker ships or pipelines.

Traditionally, pipelines and oceangoing tankers have delivered the vast majority of crude to U.S. refineries, accounting for approximately 93% of total receipts (in barrels) in 2012. Although other modes of transportation—rail, barge, and truck—have accounted for a relatively minor portion of crude oil shipments, volumes have been rising very rapidly. The volume of crude oil carried by rail increased 423% between 2011 and 2012, and the volume moving by barge, on inland waterways as well as along intracoastal routes, increased by 53%. The volume of crude oil shipped by truck rose 38%

between 2011 and 2012. **Figure 2** shows the change in transportation by mode between 2008 and 2012.

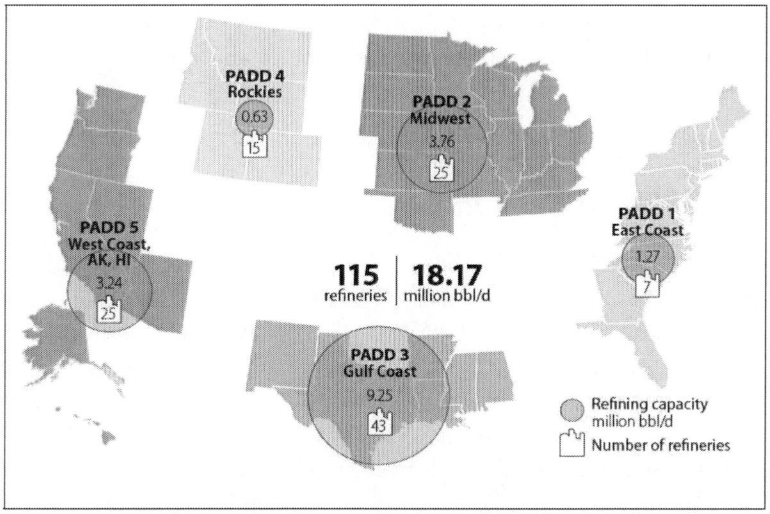

Sources: Congressional Research Service; Energy Information Administration.

Note: PADD = Petroleum Administration for Defense Districts, five districts established by executive order during World War II for gasoline rationing.

Figure 1. U.S. Refinery Capacity by PADD in 2012.

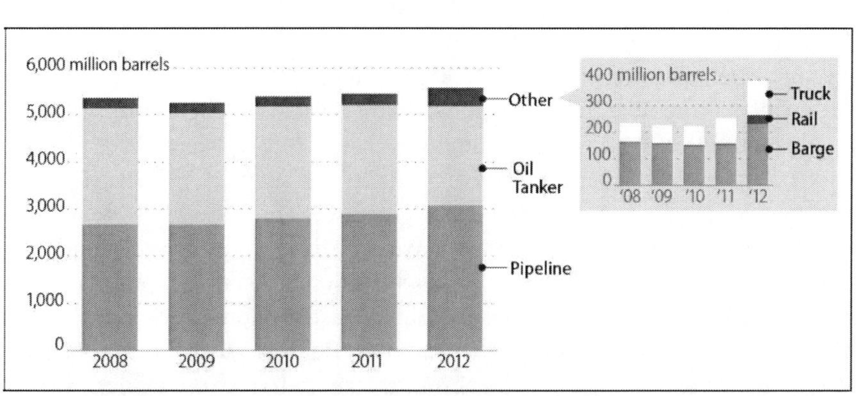

Source: Prepared by CRS; data from EIA, *Refinery Capacity Report*, Table 9, June 2013.

Notes: Some shipments may involve multiple modes, such as rail to barge. This figure indicates only the mode used for the last leg of such shipments.

Figure 2. U.S. Refinery Receipts of Crude Oil by Mode of Transportation.

Rail is a relatively high-cost method of transporting oil. Although crude oil transportation costs are typically not a major driver of refiner profitability, refiners are typically wary of incurring any costs that are higher than those faced by their competitors, as all refined petroleum products sold in a region tend to command the same price independent of the refinery that produced them.

The Economics of Oil by Rail

In the short run, rapid expansion of oil production in the Bakken—production volumes increased nearly ten-fold between 2005 and 2013[11]—strained the capacity of existing pipelines and of refiners able to process the oil. Finding ready buyers was difficult, resulting in discounted prices compared to other crude oil traded in the U.S. market. With Bakken crude selling for approximately $4 to $28 per barrel less than West Texas Intermediate (WTI) crude, the U.S. reference price for crude grade, refiners found it profitable to utilize the North Dakota oil delivered by rail even though the rail transportation cost is perhaps $5 to $10 per barrel higher than pipeline costs.

Rail has also been critical to development of Canadian oil sands. Although the vast majority of crude oil imports from Canada are delivered via existing pipeline, imports by rail are estimated to have increased from 1.6 million barrels in 2011 to 40 million barrels in 2013. Construction of the proposed Keystone XL pipeline could move a significant proportion of these shipments off the rails, as pipeline transportation is likely to cost less per barrel.[12]

For certain refiners, the economics of using rail to transport Bakken oil supplies are even more attractive. In 2012, several refineries in the Philadelphia area were scheduled for closure. The refineries were using imported crudes, largely sourced from West Africa, which sold at a premium to WTI,[13] making their refined products, notably gasoline, uncompetitive against similar products produced by Gulf Coast refineries that used cheaper heavy crudes. By using supplies from the Bakken, these refineries have lowered their costs and have become more competitive. New owners are now investing in the refineries, including installation of high-speed rail unloaders that would allow them to use 230,000 barrels per day of Bakken crude oil by early 2014.[14] These innovations would also reduce the cost of rail transportation per barrel.

The attractiveness of rail transportation of oil may be temporary. Transporting Bakken crude by rail became cost-effective because of the price discounts created by pipeline bottlenecks. If additional oil pipeline capacity were constructed, say from North Dakota to the East Coast market, refiners would likely prefer lower-cost pipeline transportation. And if the refineries could obtain Bakken crude by pipeline, demand would increase, likely reducing or eliminating the current price discount. Without the price discount, Bakken oil would not be competitive in refining when transported by rail. On the other hand, a rising Bakken crude oil price would likely lead to greater drilling activity in the Bakken fields. Given the uncertainty about the future value of the oil and the longevity of the deposits, it is not certain that investors will undertake construction of pipelines from the Bakken fields to the East Coast. In that case, large volumes of crude could be transported by rail well into the future.

Railroads are a viable alternative to pipeline transportation largely because they offer greater flexibility. The nation's railroad network is more geographically extensive than the oil pipeline network, and better able to ship crude oil from new areas of production to North American refineries. While there are about 57,000 miles of crude oil pipeline in the United States, there are nearly 140,000 miles of railroad.[15]

The geographic flexibility of the railroad network compared to the oil pipeline network can be especially beneficial for a domestic market in flux. Railroads can increase capacity relatively cheaply and quickly by upgrading tracks and roadbeds to accommodate higher train speeds, building passing sidings or parallel tracks, increasing the frequency of switchovers from one track to the other, and upgrading signal systems to reduce the headway needed between trains. Although railroads need approval from the federal Surface Transportation Board (STB) to build new lines, they do not require STB approval to make improvements to existing lines. And even without capacity improvements, railroads can offer routings not served by pipelines.

A significant fall-off in railroad coal movements has increased railroads' capacity to transport oil over some routes. In 2013, railroads carried about 395,000 more tank cars of crude than in 2005, but about 1.3 million fewer cars of coal. To put the increase in crude traffic in perspective, crude oil represented less than 1% of total rail carloads in 2012. In the first three quarters of 2013, crude carloads increased to 1.4% of total rail car loadings.

THE U.S. RAILROAD INDUSTRY IN BRIEF

The U.S. rail network comprises seven large (Class I) railroads, which focus on moving products between North American regions. These railroads generally market to large volume, long-distance shippers. There are also roughly 500 "shortline" (Class II or III) railroads that sometimes serve as the first or final leg of a Class I rail shipment. Shortlines were often spun-off from Class I railroads because of insufficient business over the line. Class I railroads account for about 70% of system mileage, 90% of railroad employees, and about 95% of freight railroad revenue. Since crude oil movements involve non-traditional rail origins (drilling sites) and destinations (refineries), shortlines are often involved in these movements.

Railroad track is categorized into classes that determine the allowable speeds over the track.[16] Most track with the lowest speed limits is the property of shortlines. If track needs maintenance work, a railroad will issue a "slow order" on that section of track, reducing train speeds. Class I railroads have transitioned to using bigger and heavier cars, raising the maximum weight on many track sections from 263,000 lbs. to 286,000 lbs. Shortline railroads that interchange traffic with Class I railroads have had to improve their roadbeds to accommodate the heavier cars.

The railroad industry, since 1980, is mostly economically deregulated. The Surface Transportation Board can review the reasonableness of railroad rates and service in situations where the railroad is determined to have "market dominance," generally where a shipper is served by only one railroad and cannot ship economically by other means. As "common carriers," railroads are required to provide rail service upon reasonable request. Railroads do not require a special federal permit to transport crude oil. Federal railroad law preempts state and local authority, which is generally restricted to a state or local government's "police powers."

Railroad transport reportedly costs in the neighborhood of $10 to $15 per barrel compared with $5 per barrel for pipeline. In return, railroads offer oil producers certain advantages. Heated railroad tank cars improve the viscosity of oil sands crude so that less diluent needs to be added than if the product were being moved by pipeline. Generally, railroads are more willing to enter into shorter-term contracts with shippers than pipelines (one to two years versus 10 to 15 years), offering more flexibility in a rapidly changing oil market. Moving oil by train from North Dakota to the Gulf Coast or Atlantic Coast requires about five to seven days' transit, versus about 40 days for oil moving by pipeline, reducing producers' need for working capital to cover the cost of oil in transit.[17]

Crude oil often moves by unit train, a train that carries just one type of cargo in a single type of car and serving a single destination. Unit trains do not need to be switched or shunted in rail yards, saving time and reducing costs, and return to their point of origin as soon as they have been unloaded. A train consisting of 70 to 120 tank cars can carry in the neighborhood of 50,000 to 90,000 barrels of oil, depending on the type of crude.

One hindrance to the expansion of crude-by-rail has been the lack of tank cars and loading and unloading infrastructure. Much of this investment is being made by the oil industry or by rail equipment leasing companies, not railroads. As of summer 2013, manufacturers had more than 60,000 tank cars of all types on order, representing more than two years of production; the number intended for crude oil transport is unknown, but approximately 92,000 existing tank cars can be used to transport crude oil.[18] Rail terminal capacity is expected to increase fourfold from 2012 to 2015.[19] Matching the daily throughput volume of a pipeline requires several trains per day, with each train taking 13 to 24 hours to unload; oil rail terminals therefore require large areas for parallel loop tracks where multiple trains can await unloading.

Pipelines generally provide more reliable service than railroads. Among other differences, rail shipments are more affected by weather. In addition, railroads generally experience peak demand during the fall due to the grain harvest and retailers' holiday shipments. This may cause locomotives and track capacity to be in shorter supply at certain times of the year.

The Role of Barges and Ships in Domestic Crude Transportation

Many refineries traditionally have received crude from overseas and thus are located near the coastline with access to dock facilities. Some are not equipped to receive crude by rail. Hence, some railroads are transferring oil to barges for the last leg of the trip to refineries, especially in the South and Midwest. Locations where railroads transfer crude oil to barges include St. Louis and Hayti, MO; Osceola, AR; Hennepin, IL; Albany, NY; and Anacortes and Vancouver, WA. In addition, crude produced at Eagle Ford,

TX, which is located near ports, is being moved along the coast by either barge or ship.

One river barge can hold 10,000 to 30,000 barrels of oil. Two to three river barges are typically tied together in a single tow that carries 20,000 to 90,000 barrels, about the same load as a unit train. Coastal tank barges designed for open seas, known as articulated tug-barges, or ATBs,[20] can hold 50,000 to 185,000 barrels, although newer ATBs can carry as much as 340,000 barrels, comparable to the capacity of coastal tankers. Much larger crude oil tankers are used to move Alaska oil to West Coast refineries.

The Jones Act

The Jones Act may have a profound impact on where crude oil is sourced and how it is transported. The Jones Act requires that vessels transporting cargo between two U.S. points be built in the United States, as well as crewed and at least 75% owned by U.S. citizens.[21] The domestic build requirement for tanker ships, in particular, has been identified as contributing to higher costs in moving domestic crude oil along the coasts.[22] Domestically built tankers are about four times the price of foreign-built tankers,[23] and there is limited capacity in U.S. shipyards to build them. Much of the existing crude oil tanker fleet was built since 2000 to meet Oil Pollution Act of 1990 (P.L. 101-380) requirements that tankers calling at U.S. ports have double hulls. Two crude carriers are expected to be delivered in 2014 to replace two vessels in Alaska trade.

As of June 2013, the Jones Act-eligible fleet of crude oil tankers consisted of 10 ships, all employed in moving Alaska crude oil to the U.S. West Coast or to a refinery in Alaska.[24] Since annual Alaska oil production has fallen by about 46% over the last decade, the Jones Act crude oil fleet has been in decline. About 30 Jones Act-eligible tankers carry chemicals or refined petroleum products, such as gasoline or jet fuel, but these ships do not readily alternate between carrying dirty oil (crude oil, residual fuel oil, asphalt) and refined (clean) petroleum products because the tanks would have to be extensively washed after carrying dirty product, a time - consuming and costly process. Some product vessels have fundamentally different designs from crude carriers and would require a layup in a shipyard to be converted to move crude oil.

In the past, Jones Act tanker shipping rates have generally been higher than, but have largely moved in tandem with, the rates for tankers coming from overseas. Phillips 66 recently chartered two Jones Act tankers to move crude oil from Eagle Ford, TX, to a refinery in Linden, NJ,[25] but it remains to be seen whether other refiners will deem tankers cost-competitive with trains and barges in moving crude oil from the Gulf Coast to the Northeast.

The Role of Tank Trucks

Tank trucks operating on U.S. roadways have been an important link in moving crude oil from domestic drilling sites to pipelines and rail terminals. A typical tank truck can hold 200 to 250 barrels of crude oil. Trucks readily serve the need for gathering product, as the hydraulic fracturing method of drilling employed in tight oil production involves multiple drilling sites in an area and the location of active wells is constantly in flux. A large volume of crude oil is being transported by truck between production areas and refineries in Texas because of the close proximity of the two.

OIL SPILL CONCERNS

Each mode of oil transportation—pipelines, vessels, rail, and tanker trucks—involve some risk of oil spills. Over the period 1996-2007, railroads consistently spilled less crude oil per ton-mile than trucks or pipelines. Barges and domestic tanker ships have much lower spillage rates than trains (**Figure 3**). However, the data in **Figure 3** precede the recent dramatic increase in oil transportation by rail.

Given the comparatively small capacity of a rail tank car, around 700 barrels, the total amount spilled from even a major derailment is likely to be small compared to the 260,000 barrels discharged in the 1989 grounding of the *Exxon Valdez* in Prince William Sound, AK, or the approximately 40,000 barrels discharged in the largest U.S. pipeline oil spill CRS can document, which occurred in 1991 near Grand Rapids, MN.[26] Nonetheless, spill volume is arguably a relatively unimportant factor in terms of impacts and cleanup costs. Location matters more: a major spill away from shore will likely cost considerably less to abate than a minor spill in a populated location or sensitive ecosystem. Depending on timing and location, even a small spill can cause significant harm to individual organisms and entire populations.[27]

CRS is not aware of any database that tracks oil spills from rail transport. Although spillage per ton-mile of oil transported by rail declined over time, a recent series of major accidents (**text box**) has heightened concern about the risks involved in shipping crude by rail.

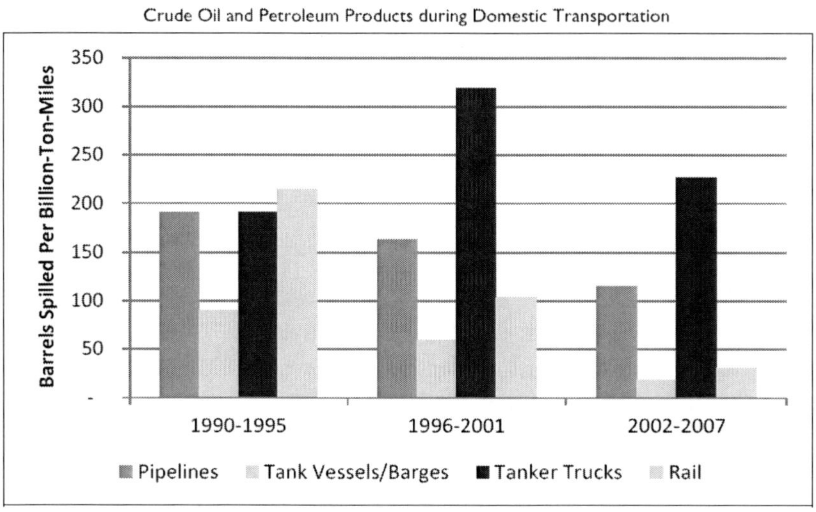

Source: Prepared by CRS; oil spill volume data from Dagmar Etkin, *Analysis of U.S. Oil Spillage*, API Publication 356, August 2009; ton-mile data from Association of Oil Pipelines, *Report on Shifts in Petroleum Transportation: 1990-2009*, February 2012.

Notes: Pipelines include onshore and offshore pipelines. The time periods were chosen based on the available annual data for both spill volume and ton-miles. The values for each time period are averages of annual data for each six-year period.

Figure 3. Oil Spill Volume per Billion-Ton-Miles.

OIL BY RAIL DERAILMENTS IN 2013 AND 2014

Lac Mégantic, Quebec—On July 5, 2013, a train with 72 loaded tank cars of crude oil from North Dakota moving from Montreal, Quebec, to St. John, New Brunswick stopped at Nantes, Quebec, at 11:00 pm. The operator and sole railroad employee aboard the train secured it and departed, leaving the train on shortline track with a descending grade of about 1.2%. At about 1:00 AM, it appears the train began rolling down the descending grade toward the town of Lac-Mégantic, about 30 miles from the U.S. border. Near the center of town, 63 tank cars derailed, resulting in multiple explosions and subsequent fires. There were 47 fatalities and extensive damage to the town. 2,000 people were evacuated. The initial determination was that the braking force applied to the train was insufficient to hold it on the 1.2% grade and that the crude oil released was more volatile than expected.

Gainford, Alberta—On October 19, 2013, nine tank cars of propane and four tank cars of crude oil from Canada derailed as a Canadian National train was entering a siding at 22 miles per hour. About 100 residents were evacuated. Three of the propane cars burned, but the tank cars carrying oil were pushed away and did not burn. No one was injured or killed. The cause of the derailment is under investigation.

Aliceville, Alabama—On November 8, 2013, a train hauling 90 cars of crude oil from North Dakota to a refinery near Mobile, AL, derailed on a section of track through a wetland near Aliceville, AL. Thirty tank cars derailed and some dozen of these burned. No one was injured or killed. The derailment occurred on a shortline railroad's track that had been inspected a few days earlier. The train was travelling under the speed limit for this track. The cause of the derailment is under investigation.

Casselton, North Dakota—On December 30, 2013, an eastbound BNSF Railway train hauling 106 tank cars of crude oil struck a westbound train carrying grain that shortly before had derailed onto the eastbound track. Some 34 cars from both trains derailed, including 20 cars carrying crude, which exploded and burned for over 24 hours. About 1,400 residents of Casselton were evacuated but no injuries were reported. The cause of the derailments and subsequent fire is under investigation.

Plaster Rock, New Brunswick—On January 7, 2014, 17 cars of a mixed train hauling crude oil, propane, and other goods derailed likely due to a sudden wheel or axle failure. Five tank cars carrying crude oil caught fire and exploded. The train reportedly was delivering crude from Manitoba and Alberta to the Irving Oil refinery in Saint John, New Brunswick. About 45 homes were evacuated but no injuries were reported.

Philadelphia, Pennsylvania—On January 20, 2014, seven cars of a 101-car CSX train, including six carrying crude oil, derailed on a bridge over the Schuylkill River. No injuries and no leakage were reported, but press photographs showed two cars, one a tanker, leaning over the river.

In March and April 2013, there were two derailments of Canadian Pacific trains, one in western Minnesota and the other in Ontario, Canada; less than a tank car of oil leaked in each derailment and neither incident caused a fire.

The increasing deployment of unit trains changes the risks involved in shipping oil by rail in two ways. Unit trains of crude oil concentrate a large amount of potentially environmentally harmful and flammable material, increasing the probability that, should an accident occur, large fires and explosions could result. This risk is similar to that of unit trains carrying ethanol, and maybe greater than that of mixed freight trains in which various hazardous materials, such as explosives and toxic-by-inhalation materials, are sequenced among other cars according to federal regulations.[28] On the other hand, while unit trains concentrate a voluminous quantity of potentially dangerous material, they may offer safety benefits from avoiding the decoupling and re-coupling of cars in rail yards, which involve high-impact forces and introduce opportunity for human error.

Special Concerns about Canadian Dilbit

Oil companies generate substantial quantities of crude oil and related substances from the natural bitumen in oil sands, particularly deposits in Alberta, Canada. In 2012, the United States imported 438 million barrels of oil sands-derived crude oils, 125% more than in 2005.[29] Because bitumen is highly viscous, it is transported mostly in the form of diluted bitumen, or dilbit, containing naptha or other materials that make it flow more easily.

Some commenters have argued that due to its physical characteristics, dilbit presents greater risks of oil spills than conventional crude, with potentially greater impacts to the environment.[30] Other stakeholders and organizations have questioned these conclusions.[31] A study released by the National Research Council in 2013, conducted at the direction of Congress,[32] found that the characteristics of dilbit do not increase the likelihood of spills.[33] The extent to which these findings are applicable to rail transport of crude is open to debate, as rail tanker cars may have different operating parameters (e.g., temperature) and physical standards (e.g., wall thickness), or may transport different forms of oil sands-derived crude oil, decreasing the relevance of the NRC findings.

However, observations in the aftermath of a 2010 pipeline spill are consistent with the assertion that dilbit may pose different hazards, and possibly different risks, than other forms of crude oil. On July 26, 2010, a pipeline owned by Enbridge Inc. released approximately 850,000 gallons of dilbit into Talmadge Creek, a waterway that flows into the Kalamazoo River in Michigan.[34] Three years after the spill, response activities continued,[35] because, according to EPA, the oil sands crude "will not appreciably biodegrade."[36] The dilbit sank to the river bottom, where it mixed with sediment, and EPA has ordered Enbridge to dredge the river to remove the oiled sediment.[37] As a result of this order, Enbridge estimated in September 2013 its response costs would be approximately $1.035 billion,[38] which is substantially higher than the average cost of cleaning up a similar amount of conventional oil.[39]

Special Concerns about Bakken Crude

The properties of Bakken shale oil are highly variable, even within the same oil field. In general, however, Bakken crude oil is much more volatile than other types of crude.[40] Its higher volatility may have important safety implications.

In January 2014, the Pipeline and Hazardous Materials Safety Administration (PHMSA) within the Department of Transportation (DOT) issued a safety alert warning that recent derailments and resulting fires indicate that crude oil being transported from the Bakken region may be more flammable than traditional heavy crude oil.[41] PHMSA, whose rules are enforced by the Federal Railroad Administration with respect to railroads, reinforced the requirement to properly test, characterize, classify, and where appropriate sufficiently degasify hazardous materials prior to and during transportation. Under its initiative "Operation Classification," PHMSA is to continue to collect samples and measure the characteristics of Bakken crude as well as oil from other locations.

Federal Oversight of Oil Transport by Rail

The Federal Railroad Administration (FRA) has jurisdiction over railroad safety. It has about 400 federal inspectors throughout the country and also utilizes state railroad safety inspectors. State inspectors predominantly enforce federal requirements because federal rail safety law preempts state law, and federal law is pervasive. The FRA uses past incident data to determine where its inspection activity should be targeted, although the FRA Administrator recently stated that in light of the growth of crude-by-rail transportation, the agency also must look for "pockets of risk."[42] FRA regulations cover the safety of track, grade crossings, rail equipment, operating practices, and movement of hazardous materials (hazmat). The Pipeline and Hazardous Materials Safety Administration within DOT (PHMSA) issues requirements for the safe transport of hazmat by all modes of transportation, which the FRA enforces with respect to railroads.[43]

Rail incidents are investigated by the National Transportation Safety Board (NTSB), an independent federal agency. The NTSB makes recommendations toward preventing future incidents based on its findings. Unlike the FRA, the NTSB is not required to weigh the costs against the benefits when considering additional safety measures and it has no regulatory authority. Many of the NTSB's recommendations concerning oil transport by rail are identical to those it previously issued for transporting ethanol by rail. While the FRA has largely agreed with NTSB's recommendations, its rulemaking process involves consultation with industry advisory committees, and it must determine which of the many rail safety measures under evaluation deserve priority. Implementing a change in FRA regulations can take years.

U.S. safety requirements apply to any train operating in the United States, regardless of its origin or destination. Canadian safety regulations are very similar but do not exactly mirror U.S. requirements. Cross-border shipments must meet the requirements of both countries. Safety standards established by the rail industry, which often exceed government requirements, apply to both U.S. and Canadian railroads.

When a rail incident results in the release of oil, state, territorial, or local officials are typically the first government representatives to arrive at the scene and initiate immediate safety measures to protect the public. The National Oil and Hazardous Substances Pollution Contingency Plan, often referred to as the National Contingency Plan (NCP), indicates that state, territorial, or local officials may be responsible for conducting evacuations of affected populations. These first responders also may notify the National Response Center to elevate an incident for federal involvement, at which point the coordinating framework of the NCP would be applied.

Unlike most federal emergency response plans, which are administrative mechanisms, the NCP is codified in federal regulation and is binding and enforceable.[44] The NCP regulations apply to applicable spills from vessels, pipelines, onshore facilities, and offshore facilities. The definition of "onshore facility" includes, but is not limited to "motor vehicles and rolling stock."[45]

If an oil discharge affects navigable waterways, shorelines, or "natural resources belonging to, appertaining to, or under the exclusive management authority of the United States,"[46] Section 311 of the Clean Water Act, as amended by the Oil Pollution Act of 1990, Section 311(c), provides explicit federal authority to respond.[47] The term "discharge" is defined broadly and is not linked to specific sources of oil. The President has the authority to perform cleanup immediately using federal resources, monitor the response efforts of the spiller, or direct the spiller's cleanup activities.[48] Several executive orders have delegated the President's response authority to the Environmental Protection Agency (EPA) within the "inland zone" and to the U.S. Coast Guard within the coastal zone, unless the two agencies agree otherwise.[49] The lead federal agency serves as the On-Scene Coordinator to direct the federal resources used in a federal response.

Regulations require that railroads have either a so-called "basic" response plan or a more "comprehensive" response plan, depending on the volume capacity of the rail car transporting the oil.[50] Comprehensive plans are subject to FRA approval, and must ensure by contract or other means that personnel and equipment are able to handle a worst-case discharge.[51] However, the regulatory threshold for the comprehensive response plan is a tank car holding more than 1,000 barrels, so does not apply to the DOT-111 tank cars used today, which hold around 700 barrels of oil apiece. For these smaller tank cars, railroads must prepare only "basic" response plans, which are not subject to FRA approval.

This threshold was established in 1996,[52] before the advent of oil unit trains, each of which may transport, in aggregate, approximately 70,000 barrels (almost 3 million gallons) of oil. The NTSB recently recommended that the threshold for comprehensive plans be lowered to take into account the use of unit trains.[53]

ISSUES FOR CONGRESS

While oil by rail has demonstrated benefits with respect to the efficient movement of oil from producing regions to market hubs, the dramatic increase in oil by rail shipments has generated interest in several related issues. These include railroad safety,[54] environmental concerns, and trade-offs over rail versus pipeline development.

Railroad Safety and Incident Response

Prior to the Lac Mégantic derailment, the FRA had increased its inspection activity with regard to trains carrying crude oil. After the derailment, the FRA and PHMSA (along with Transport Canada) initiated a comprehensive review of safety requirements.[55] Three areas of active discussion involve tank car design, prevention of derailments, and railroad operations. On January 16, 2014 U.S. DOT officials met with railroads and oil shippers and announced that within 30 days the industry would submit a plan to improve oil-by-rail safety covering trains speeds, routing protocols, and tank car design.[56]

Tank Car Safety Design

DOT establishes construction standards for tank cars.[57] A tank car used for oil transport is roughly 60 feet long, about 11 feet wide, and 16 feet high (see **Figure 4**). It weighs 80,000 pounds empty and 286,000 pounds when full. It can hold about 30,000 gallons or 715 barrels of oil, depending on the oil's density. The tank is made of steel plate, 7/16 of an inch thick (see 49 CFR §179.201).[58] An oil tank car is typically loaded from the top valve and unloaded from the bottom valve. Loading or unloading each car may take several hours, but multiple cars in a train can be loaded or unloaded simultaneously.

In some incidents, oil has been released from the ends of tank cars because the coupler from a neighboring car has punctured the tank during derailment. Valves at the top and bottom of the cars have also been sheared off or otherwise opened during derailment. Efforts to improve crashworthiness have focused on reinforcing the shells of tank cars at both ends or adding protective shields ("jackets"), modifying couplers to prevent decoupling, adding skid protection or diversion shields to protruding valves, eliminating or modifying bottom valves, and increasing insulation for fire protection.[59]

The FRA and PHMSA have questioned whether Bakken crude oil, given its characteristics, would more properly be carried in tank cars that have additional safety features, such as those found on pressurized tank cars used for hauling explosive liquids.[60] Some of these features add weight to the car, thus possibly increasing the number of shipments needed to move a given amount of product. The railroad industry established additional standards in October 2011 for newly built cars that address some but not all of the safety features that FRA and PHMSA are considering.

Rail cars have an economic life of 30 to 40 years, so conversion of the fleet to a new car standard could take some time. DOT has asked for further information on the costs and benefits of retrofitting the existing fleet.[61] In

November 2013, the Association of American Railroads stated it supports either retrofitting or phasing out oil tank cars built before October 2011 (a fleet of about 78,000 cars) and modifying those built after October 2011 (about 14,000 cars).[62] Some Members of Congress have urged DOT to expedite the rulemaking process concerning new tank car safety requirements.[63]

Preventing Derailments

An analysis of freight train derailments from 2001-2010 on Class I railroads' mainline track found that broken rails or track welds were the leading cause of derailments, by far.[64] These problems caused 670 derailments over the period, while the next leading problem (track geometry defects) caused just over 300 derailments. Broken rails or welds also resulted in more severe incidents, derailing an average of 13 railroad cars instead of 8.6 cars for all other causes. Broken rails or welds accounted for 23% of total cars derailed. A separate study covering the same time period found that track problems were the most important causes of derailments, followed by problems with train equipment.[65]

In the Rail Safety Improvement Act of 2008 (P.L. 110-432, Section 403(a)), Congress requested that the FRA study and consider revising the frequency and methods of track inspection. FRA conducted the study and on January 24, 2014 issued a final rule on improving rail integrity.[66] The new rule requires railroads to achieve a specified track failure rate rather than scheduling inspections based on the calendar or traffic volume. It also allows railroads to maximize use of rail inspection vehicle time by prioritizing remedial action when track defects are detected.

The railroad industry takes a number of extra safety precautions for trains carrying certain amounts and kinds of hazardous materials (referred to as "key" trains).[67] Key trains include unit trains of crude oil. In response to the Lac Mégantic derailment, the industry recently modified the guidelines for key trains to include

- restricting train speeds to less than 50 mph;[68]
- increasing the frequency of track maintenance;
- installing wayside defective equipment detectors, such as "hot box" detectors that detect wheels with faulty bearings, every 40 miles, with specific protocols for conductors when defects are indicated;
- using only track in good enough condition to support speeds of 25 mph or higher.

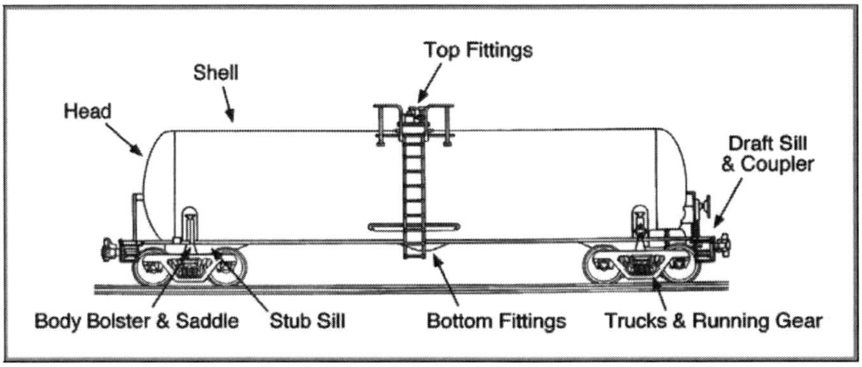

Figure 4. Non-jacketed, Non-pressure Tank Car.

Reducing train speed can reduce the number of cars that derail as well as the likelihood that product will be released from those tank cars.[69]

Shortline Track

It is often the case that a Class I railroad, prior to turning over the operation of a line to a shortline, did not maintain it to the same standards as its busy mainlines. Shippers using a shortline often do not require higher-speed track because they ship infrequently or because the commodities they ship are not time-sensitive. Thus, shortline track is frequently maintained at a lower standard than Class I railroads' track. The Lac Mégantic, Quebec, and Aliceville, AL, crude oil derailments occurred on shortline track. Members of Congress have been concerned with preserving shortline rail service, reflected in a federal loan program for track rehabilitation and improvement and a tax credit for shortline track maintenance.[70]

Railroad Operations

A number of specific operational issues have been found relevant to railroad safety, in general, or to oil by rail transportation specifically.

Terminal Operations

In September, 2013, the FRA solicited public comment on whether current regulations concerning transfer of crude oil from and to tank cars are adequate considering recent practices at transload facilities. Its request for public comment asked for information about what entity controls trains on loop tracks at rail loading terminals and what procedures have been adopted to prevent unintended movement during loading.[71]

Railroad Crew Size

Following the Lac Mégantic disaster, legislation (H.R. 3040) was introduced in Congress to require two-person crews on all trains. In the United States, the FRA does not specify in regulation how many persons must operate a train, but notes that the various tasks required while a train is moving essentially necessitate at least a two-person crew. Most trains operate with an engineer and a conductor, but some shortline railroads may operate trains with a single crew member. The FRA appears to be moving toward a regulation requiring two-person crews while allowing for some exceptions.[72] One potential trade-off is that distraction by a fellow crew member has been found to be a factor in past accidents.

Positive Train Control

Railroads are in the process of implementing positive train control (PTC), a system that is designed to override human error in controlling the speed and movement of trains. Congress required that this system be installed on routes carrying passengers or poison- or toxic-by-inhalation hazardous materials (Section 104 of P.L. 110-432), a requirement that applies to about 60,000 miles of railroad. Current law does not require installation of PTC solely because a track carries crude oil, but the law authorizes the FRA to expand the scope of tracks required to have PTC. PTC is not required on track in or near rail yards. The cost and timeline for implementing PTC are topics of current debate among policymakers and stakeholders.[73]

Route Selection

In the Implementing Recommendations of the 9/11 Commission Act of 2007 (P.L. 110-53, Section 1551), Congress required railroads carrying certain kinds and quantities of potentially dangerous commodities to assess the safest and most secure routes for trains carrying these products and to minimize delays and storage for rail cars containing these products. These requirements currently apply to explosive, toxic-by-inhalation, and radioactive material.[74] Security regulations also require that rail cars containing these commodities not be left unattended when being transferred from one carrier to another or between carrier and shipper.[75] The law resulted from efforts by cities like Washington, DC, and Pittsburgh to ban trains carrying hazardous materials.[76]

The FRA may consider whether this routing analysis should also apply to unit trains of crude oil.[77] Such a requirement would be controversial because avoiding large urban areas can increase the length of time such trains are in transit and because smaller towns and rural areas likely have less capability to respond to emergencies than large cities.

Incident and Oil Spill Response

The increased use of rail for crude oil shipments is likely to increase the number of incidents, some of which may involve oil spills. As described above, the National Oil and Hazardous Substances Contingency Plan provides a framework for federal, state and local collaboration in response to releases of oil and hazardous substances. Considering the relative proximity of rail shipments to population centers, a potential issue for Congress is the safety and adequacy of spill response.

In addition, based on past history, increased frequency or severity of incidents related to shipments of crude oil by rail could lead some local communities to seek additional funding to ensure adequate spill response capabilities, including personnel, training, equipment, and community notification.

The Accuracy of Train Cargo Information

Crude oil may also be carried by "mixed trains"—trains carrying a variety of different commodities. With mixed trains, it is important to first responders that they have an accurate list of which cars contain what commodities (the train "consist"). Often the sequencing of cars changes en route, so the consist information provided by the crew at the scene of an incident may no longer be accurate. Although all vehicles containing hazardous materials must display placards indicating their potential dangerous characteristics (e.g., flammable, corrosive, explosive), responders often need more specific information about the commodities involved in an incident. One potential remedy under consideration is an electronic manifest system that would offer the capability of easier updates. In MAP-21, Congress authorized PHMSA to conduct pilot projects on paperless hazmat information sharing among carriers (of various modes including rail) and first responders.[78] A potential drawback raised by the railroads is that electronic devices at the scene of an incident could encounter technical problems. Another remedy is greater diligence by railroad crew in keeping the paper consist up to date. The NTSB has asked whether a copy of the consist should also be kept at the end of a train in case the copy kept by the crew at the head of the train is lost in an incident.

Rail vs. Pipeline Development

Certain rail routings of crude oil could be replaced by reconfiguring the existing pipeline network and constructing additional pipeline capacity. In general, pipelines could provide safer, less expensive transportation than railroads, assuming that pipeline developers are able to assure markets for the oil they hope to carry.

Pipeline development could be particularly important for shipments of crude oil from Canada to the United States. In light of growing Canadian exports, several proposals have been made to expand the cross-border pipeline infrastructure. Of the five major pipelines currently linking Canadian petroleum producing regions to markets in the United States, two (Alberta Clipper and Keystone) began service in 2010. A permit application for a sixth pipeline, Keystone XL, a very large project which would also transport some Bakken crude, was initially submitted in 2008 and is in the final stages of review by the U.S. Department of State.[79] Keystone XL has been the subject of intense scrutiny and debate by Congress, the Executive Branch, and numerous stakeholders. The Keystone XL review and approval process is highly contested, and the pipeline's approval remains uncertain.

Other proposed oil pipeline projects, such as the reversal of the Portland-Montreal oil pipeline to enable export of Canadian oil via a marine terminal in Maine, are also encountering greater public scrutiny and opposition. On the whole, the barriers to new oil pipeline approval in any jurisdiction seem to have risen significantly since Alberta Clipper and Keystone were completed.

Shipment of oil by rail is, in many cases, an alternative to new pipeline development. This involves tradeoffs in terms of both transportation capacity and safety. In its ongoing review of the Keystone XL pipeline proposal, the State Department has argued that, if the pipeline is not constructed, additional oil-by-rail capacity will be developed instead. As the State Department's 2014 Final Environmental Impact Statement for the Keystone XL project states,

> In the past 2 years, there has been exponential growth in the use of rail to transport crude oil throughout North America, primarily originating from the Bakken in North Dakota and Montana, but also

> increasingly utilized in other production areas, including the [Western Canadian Sedimentary Basin]. Because of the flexibility of rail delivery points, once loaded onto trains the crude oil could be delivered to refineries, terminals, and/or port facilities throughout North America, including the Gulf Coast area.[80]

Consistent with this view, both Canadian National Railway and Canadian Pacific Railway reportedly have been pursuing a "pipeline on rails" business strategy, including new track investments, to move Canadian crudes to new markets throughout North America.[81] Increasing cross-border movements of crude oil by rail on existing track does not require State Department approval, so such an approach seeks to avoid regulatory delays. While the potential volumes associated with rail transportation of crude could be lower than pipeline volumes, they could still be significant. Some analysts have suggested that oil-by-rail volumes could be large enough to make a major new pipeline project like Keystone XL unnecessary.[82] Similar arguments could apply to other oil transportation corridors within North America.

Others are less certain that oil by rail can substitute so readily for pipeline capacity, as rail expansion would require significant infrastructure development including loading and unloading facilities, track capacity, and, possibly, additional tank car availability. The State Department's analysis finds that under certain conditions, including particular oil and oil transportation prices, "there could be a substantial impact on oil sands production levels."[83] Other market analyses similarly find that in the short and medium term some production could be curtailed.[84]

Refiner economics may ultimately favor pipelines over rail, although those investment decisions will be determined by market forces. When it comes to safety, however, the federal government plays a major role, and thus may have considerable influence on infrastructure expansion. Some participants in the Keystone XL debate, for example, have asserted that recent oil-by-rail incidents underscore the need for a new pipeline as, in their view, a safer mode of transportation for Canadian crudes,[85] while others insist that safety comparisons between the two transportation modes are less conclusive.[86] On balance, however, it seems likely that policies that raise the cost of transporting oil by rail would increase the attractiveness of pipeline development, and, for that matter, expansion of crude oil transportation by barges, tanker ships, and tanker trucks.

Rail Transport and Crude Oil Exports

The large increase in U.S. oil production has led some Members of Congress to advocate changing the law that generally prohibits exports of crude oil.[87] An increase in crude oil exports would likely require greater use of rail transportation, as the crude oil pipeline network is not oriented to serve export ports. Some environmental groups have stated their opposition to construction of new rail facilities or terminals that would facilitate oil exports, as they believe increased exports will encourage environmentally damaging production in the United States and Canada.[88]

End Notes

[1] The terms "oil sands" and "tar sands" are often used interchangeably to describe a particular type of nonconventional oil deposit. Opponents of the resource's development often use the term "tar sands," which arguably carries a negative connotation; proponents typically refer to the material as oil sands. The use of this term is not intended to reflect a point of view, but to adopt the term most commonly used by the primary executive-branch agencies involved in recent oil sands policy issues.

[2] Edward R. Hamberger and Andrew J. Black, "Freight Rail and Pipelines Deliver Energy for America," *The Hill, Congress Blog*, November 5, 2013, http://thehill.com/blogs/congress-blog/energy-environment/189187-freight-rail-and-pipelines-deliver-energy-for-america.

[3] Estimates by CRS based on data from Dagmar Etkin, *Analysis of U.S. Oil Spillage*, API Publication 356, August 2009, and Association of Oil Pipelines, *Report on Shifts in Petroleum Transportation: 1990-2009*, February 2012.

[4] See, for example, Office of Senator John Hoeven, "Hoeven to Meet Saturday with BNSF Railway President and CEO to Address Railroad Safety," press release, January 3, 2014.

[5] Energy Information Administration, *U.S. Crude Oil Supply & Disposition*, http://www.eia.gov/dnav/pet/ pet_sum_crdsnd_k_a.htm. A barrel of oil is equal to 42 gallons.

[6] CRS Report CRS Report R43128, *Oil Sands and the Oil Spill Liability Trust Fund: The Definition of "Oil" and Related Issues for Congress*, by Jonathan L. Ramseur.

[7] "US Crude Production Tops Imports For The First Time Since 1995," *Oil Daily*, November 14, 2013.

[8] M. Tennyson, et al., *Assessment of Remaining Recoverable Oil in Selected Major Oil Fields of the Permian Basin, Texas and New Mexico*, 2012, USGS, http://pubs.usgs.gov/fs/2012/3051/.

[9] "Light" refers to oils with low specific gravity. "Sweet" refers to oils with low sulfur content. Light, sweet crudes are more valuable than heavier or sourer crude oils.

[10] For further information on the petroleum refining industry, refer to CRS Report R41478, *The U.S. Oil Refining Industry: Background in Changing Markets and Fuel Policies*, by Anthony Andrews et al.

[11] Energy Information Administration crude oil production data, by state, available at http://www.eia.doe.gov.

[12] For more information about the Keystone XL pipeline, see CRS Report R41668, *Keystone XL Pipeline Project: Key Issues*, by Paul W. Parfomak et al.

[13] Energy Information Administration price data available at http://www.eia.doe.gov.

[14] Matthew Phillips, "North Dakota's Bakken Oil Finally Hits the East Coast," *Bloomberg Businessweek*, February 6, 2013.

[15] Pipeline data from PHMSA, railroad mileage from Association of American Railroads (includes shortline rail mileage, does not include parallel trackage).

[16] See 49 C.F.R. §213.9.

[17] BB&T Capital Markets, "Examining The Crude By Barge Opportunity," June 10, 2013, p. 15.

[18] "Freight Car Market Headed for New Growth in 2014," *Railway Age,* July 29, 2013.

[19] E. Russell Braziel, RBN Energy Inc. presentation at CSIS conference, *North American Oil and Gas Infrastructure, Shale Changes Everything*," November 14, 2013.

[20] The bow of the tug fits into a notch in the stern of the barge and the tug is hinged to the barge on both sides of its hull, allowing fore and aft (pitch) movement, such as over sea swells.

[21] The law is codified at chapters 81, 121, and 551 of Title 46, United States Code.

[22] See for instance, "Oil and the Ghost of 1920," *Wall Street Journal,* September 13, 2012; Senate Committee on Energy and Natural Resources, Testimony of Faisel Khan, Managing Director, Integrated Oil and Gas Research, Citigroup. Hearing to Explore the Effects of Ongoing Changes in Domestic Oil Production, Refining and Distribution on U.S. Gasoline and Fuel Prices, July 16, 2013.

[23] U.S. Maritime Administration, Title XI Ship Financing Guarantees, Pending and Approved Loan Applications; American Petroleum Tankers S-1 SEC Filing; RS Platou Economic Research, annual and monthly reports; press releases from Teekay Tankers, Scorpio Tankers, and Euronav.

[24] U.S. Maritime Administration, U.S. Flag Privately Owned Merchant Fleet, Oceangoing Self-propelled Vessels.

[25] "Phillips 66 Charters Tankers To Bring Shale Oil To Bayway," *Argus Media,* December 13, 2012.

[26] Sources consulted include NOAA, Oil Spill Case Histories, 1967-1991, Summaries of Significant U.S. and International Spills, 1992; U.S. Coast Guard, Notable Spills in U.S. Waters, Calendar Years 1989 -2008, 2009; Dagmar Etkin, Analysis of U.S. Oil Spillage, API Publication 356, August 2009; NOAA, Incident News, at http://incidentnews.gov; EPA, Enforcement and Compliance History Online (ECHO), at http://www.epa-echo.gov/echo/index.html.

[27] National Research Council, *Oil in the Sea III: Inputs, Fates, and Effects* (Washington, DC: National Academies of Science, February 2003).

[28] These requirements are codified at 49 CFR §174.85.

[29] Data from Canada's National Energy Board. See also CRS Report R43128, *Oil Sands and the Oil Spill Liability Trust Fund: The Definition of "Oil" and Related Issues for Congress*, by Jonathan L. Ramseur.

[30] The primary vehicle for these arguments was a 2011 report from several environmental groups. See Anthony Swift et al., *Tar Sands Pipelines Safety Risks*, Joint Report by Natural Resources Defense Council, National Wildlife Federation, Pipeline Safety Trust, and Sierra Club, February 2011.

[31] See e.g., Crude Quality Inc., *Report regarding the U.S. Department of State Supplementary Draft Environmental Impact Statement*, May 2011; and Energy Resources Conservation Board, Press Release, "ERCB Addresses Statements in Natural Resources Defense Council Pipeline Safety Report," February 2011.

[32] P.L. 112-90, §16.

[33] National Research Council, *Effects of Diluted Bitumen on Crude Oil Transmission Pipelines*, 2013.

[34] National Transportation Safety Board, *Accident Report: Enbridge Incorporated Hazardous Liquid Pipeline Rupture and Release- Marshall, Michigan, July 25, 2010*, July 2012, at http://www.ntsb.gov/.

[35] For more up-to-date information, see EPA's Enbridge oil spill website at http://www.epa.gov/enbridgespill/index.html.

[36] Letter from Cynthia Giles, Environmental Protection Agency, to U.S. Department of State, April 22, 2013.

[37] EPA Removal Order, March 14, 2013, at http://www.epa.gov/enbridgespill/ar/enbridge-AR-1720.pdf.

[38] See Enbridge Inc., Third Quarter Financial Report, 2013, at http://enbridge.com/InvestorRelations/FinancialInformation/InvestorDocumentsandFilings.aspx.

[39] Based on cost estimates prepared in 2004. See Dagmar Etkin, *Modeling Oil Spill Response and Damages Costs*, Proceedings of the 5th Biennial Freshwater Spills Symposium, 2004, at http://www.environmental-research.com.

[40] Bryden, K. J., Grace Catalysts Technologies, Columbia, Maryland; Habib Jr., E. T., Grace Catalysts Technologies, Columbia, Maryland; Topete, O. A., Grace Catalysts Technologies, Houston, Texas, Processing shale oils in FCC: Challenges and opportunities 09.01.2013 http://www.hydrocarbonprocessing.com/Article/3250397/Processing-shale-oils-in-FCC-Challenges-and-opportunities.html.

[41] Pipeline and Hazardous Materials Safety Administration, Safety Alert—January 2, 2014, Preliminary Guidance from OPERATION CLASSIFICATION. This advisory is a follow-up to the PHMSA and Federal Railroad Administration (FRA) joint safety advisory published November 20, 2013 [78 FR 69745].

[42] FRA Administrator Szabo, Opening Remarks to RSAC Meeting, October. 31, 2013; http://www.fra.dot.gov/eLib/Details/L04852.

[43] FRA and PHMSA are agencies within DOT, which has the emergency authority to restrict or prohibit transportation that poses a hazard of death, personal injury, or significant harm to the environment. See 49 U.S.C. §20104.

[44] 40 C.F.R. Part 112.

[45] 40 C.F.R. §300.5. This definition is also found in the Clean Water Act and OPA.

[46] The Oil Pollution Act of 1990 expanded and clarified the President's authorities under Section 311 of the Clean Water Act (33 U.S.C. §2701 et. seq.). For a more in-depth discussion of the Oil Pollution Act, see CRS Report RL33705, *Oil Spills in U.S. Coastal Waters: Background and Governance*, by Jonathan L. Ramseur.

[47] 33 U.S.C. §1321. In addition, the Comprehensive Environmental Response, Compensation, and Liability Act (CERCLA) of 1980 expanded the authorities of the President to respond to releases of hazardous substances into the environment more broadly than CWA Section 311. See CRS Report R41039, *Comprehensive Environmental Response, Compensation, and Liability Act: A Summary of Superfund Cleanup Authorities and Related Provisions of the Act*, by David M. Bearden. For further details, see CRS Report R43251, *Oil and Chemical Spills: Federal Emergency Response Framework*, by David M. Bearden and Jonathan L. Ramseur.

[48] 33 U.S.C. §1321(c).

[49] Executive Order 12777, "Implementation of Section 311 of the Federal Water Pollution Control Act of October 18, 1972, as amended, and the Oil Pollution Act of 1990," 56 *Federal Register* 54757, October 22, 1991.

[50] 49 C.F.R. Part 130.

[51] See 49 C.F.R. §130.31(a) with 49 C.F.R. §130.31(b).

[52] 61 *Federal Register* 30541 (June 17, 1996).

[53] NTSB, Safety Recommendation R-14-4 through -6, directed to PHMSA, January 21, 2014.

[54] For instance, see letter from Senators Rockefeller and Wyden to U.S. DOT and DOE dated January 9, 2014.

[55] See FRA's Emergency Order No. 28 (78 *Federal Register* 48218), the agencies' Joint Safety Advisory published August 7, 2013 (78 *Federal Register* 48224), referral of safety issues to

FRA's Railroad Safety Advisory Committee (78 *Federal Register* 48931), and a NPRM related to rail hazmat (78 *Federal Register* 54849).

[56] "Rail, Oil Industries to Make Safety Changes for Transporting Crude," *Wall Street Journal,* January 16, 2014.

[57] The tank cars used to transport crude oil fall under DOT specification 111. See 49 C.F.R. §179.

[58] 49 C.F.R. §179.201.

[59] For a discussion of NTSB's recommendations concerning DOT-111 tank cars, in reference to the derailment of an ethanol unit train in Cherry Valley, IL, see NTSB Safety Recommendation R-12-5 through -8, March 2, 2012.

[60] Pressurized tank cars (DOT specification 105 and 112) have thicker shells and heads, metal jackets, strong protective housings for top fittings, and no bottom valves.

[61] 78 FR 54849 - 54861, September 6, 2013.

[62] For comments filed on this rulemaking see http://www.regulations.gov and search under docket no. PHMSA-2012-0082.

[63] See letter from Senator Schumer to PHMSA and FRA dated July 22, 2013 and news release by Senator Hoeven on January 15, 2013 indicating that a DOT final rule on tank cars would not be issued until after January 2015.

[64] T87.6 Task Force Summary Report, pp. 9-11; Xiang Liu, M. Rapik Saat, Christopher P.L. Barkan, "Analysis of Causes of Major Train Derailment and Their Effect on Accident Rates," *Transportation Research Record*, No. 2289, 2012, pp. 154-163.

[65] Xiang Liu, M. Rapik Saat, Christopher P.L. Barkan, *Safety Effectiveness of Integrated Risk Reduction Strategies for the Transportation of Hazardous Materials by Rail,* Paper presented at the Transportation Research Board, Annual Meeting 2013, paper no. 13-1811.

[66] 79 *Federal Register* 4234, January 24, 2014.

[67] See http://www.aar.com/cpc-1258%20ot-55-n%208-5-13.pdf.

[68] Current federal regulations (49 CFR §174.86) limit only trains carrying poisonous by inhalation materials (not crude oil) to 50 mph.

[69] Athaphon Kawprasert and Christopher P.L. Barkan, "Effect of Train Speed on Risk Analysis of Transporting Hazardous Materials by Rail," *Transportation Research Record*, No. 2159, 2010, pp. 59-68.

[70] The Railroad Rehabilitation and Improvement Financing (RRIF) program and Section 45G of the tax code.

[71] "FRA/PHMSA Additional Questions for Public Comment," Docket No. FRA-2013-0067-0016, 9/4/2013, http://www.regulations.gov.

[72] FRA presentation to Railroad Safety Advisory Committee, "Appropriate Train Crew Size Working Group Update," October 31, 2013; https://rsac.fra.dot.gov/meetings/20131031.php.

[73] For further information, see CRS Report R42637, *Positive Train Control (PTC): Overview and Policy Issues*, by John Frittelli.

[74] See 49 C.F.R. §172.820; 73 *Federal Register* 72182, November 26, 2008.

[75] See 49 C.F.R. §1580.107.

[76] *U.S. Rail News*, June 11, 2008, pp. 1-2; "Hazmat Hazards: U.S. Cities may not wait for Washington Before Trying to Reroute their own hazmat trains," *Journal of Commerce,* December 12, 2005.

[77] RSAC meeting, presentation by HAZMAT Working Group, October 31, 2013. The NTSB has recommended this change; see Safety Recommendation R-14-1 through -3, January 23, 2014.

[78] Section 33005 of P.L. 112-141.

[79] The construction, connection, operation, and maintenance of a pipeline connecting the United States with a foreign country require executive permission through a Presidential Permit under Executive Orders 11423 and 13337.

[80] U.S. Department of State, January 2014, Final EIS, Section 5.1, "No Action Alternatives."

[81] Nathan Vanderklippe, "CN, CP Push for a 'Pipeline on Rails,'" *Globe and Mail*, February 7, 2011.

[82] "Keystone Pipeline Seen as Unneeded as More Oil Moves by Rail," *CBC News*, September 10, 2013.

[83] 2014 Final EIS, p. 1.4-8.

[84] For example, Canadian Imperial Bank of Commerce, "Too Much of A Good Thing: A Deep Dive Into The North American Energy Renaissance," August 15, 2012; TD Economics, "Pipeline Expansion is a National Priority," Special Report, December 17, 2012; International Energy Agency," Medium-Term Oil Market Report," May 14, 2013.

[85] Diana Furchtgott-Roth and Kenneth P. Green, *Intermodal Safety in the Transport of Oil*, Fraser Institute, October 2013, http://www.fraserinstitute.org/uploadedFiles/fraser-ca/Content/research-news/research/publications/intermodal-safety-in-the-transport-of-oil.pdf.

[86] See, for example: Rory Johnston, "Train vs. Pipeline: What's the Safest Way to Transport Oil?" *Christian Science Monitor*, Energy Voices blog, October 22, 2013, http://www.csmonitor.com/Environment/Energy-Voices/2013/1022/Train-vs.-pipeline-What-s-the-safest-way-to-transport-oil.

[87] The Senate Energy and Natural Resources Committee held a hearing on this issue on January 30, 2014.

[88] See, for example, the comments of Sierra Club official Michael Marx in Blake Sobczak, "Environmentalists 'get real creative' to combat oil by rail," Energy Wire, January 13, 2014.

In: Transporting Crude Oil by U.S. Rail ISBN: 978-1-63117-837-5
Editor: Rosario S. McLaughlin

Chapter 2

STATEMENT OF JOSEPH C. SZABO, ADMINISTRATOR, FEDERAL RAILROAD ADMINISTRATION. HEARING ON "ENHANCING OUR RAIL SAFETY: CURRENT CHALLENGES FOR PASSENGER AND FREIGHT RAIL"*

Mr. Chairman, Ranking Member, and Members of the Subcommittee, thank you for the opportunity to appear before you today, on behalf of Secretary Foxx, to discuss the safety of our Nation's railroads. Rail is a particularly safe mode of transportation, and one that American passengers and shippers are choosing more than ever before. Today, I will first give an overview of the railroad industry's safety record and the Federal Railroad Administration's (FRA) safety program, including our implementation of the Rail Safety Improvement Act of 2008. Then, I will discuss the U.S. Department of Transportation's (DOT) actions in response to recent accidents and present FRA's vision to drive the next generation of rail safety.

FRA's mission is to enable the safe, reliable, and efficient movement of people and goods for a strong America, now and in the future. We are a data-driven agency. Every regulation, safety advisory and emergency order we issue is based on facts and sound research using advanced statistical methods

---

* This is an edited, reformatted and augmented version of a statement presented March 6, 2014 before the Senate Committee on Commerce, Science, and Transportation, Subcommittee on Surface Transportation and Merchant Marine Infrastructure, Safety, and Security.

and modeling. We closely monitor data and trends to identify, reduce, and eliminate risks.

Two straight years of record-breaking safety performance, along with significant reductions in all types of accidents since 2008, are strong evidence that FRA's approach to oversight and enforcement is effective.

THE RAILROAD INDUSTRY'S SAFETY RECORD AND FRA'S SAFETY PROGRAM

FRA's top priority is safety, and fiscal year (FY) 2012 was the safest year on record, with preliminary data from FY 2013 indicating it will be even better than FY 2012's record.

Since FY 2004:

- Total train accidents have declined by 47 percent.
- Total derailments have declined by 47 percent.
- Total highway-rail grade crossing accidents have declined by 35 percent.

These safety improvements resulted in 13-percent fewer fatalities overall (895 fatalities to 779 fatalities – 95 percent of which are trespassing or grade crossing related), 59-percent fewer employee fatalities, and 9-percent fewer injuries (9,367 injuries to 8,534 injuries) over 10 years. These improvements are impressive in their own right, but especially if you consider the regulatory workload that FRA received from the Rail Safety Improvement Act of 2008 (RSIA) *and* passenger and freight rail's growth during this same time.

- Amtrak set new ridership records in 10 of the last 11 years,
- Rail was the fastest-growing mode of public transportation, and
- Intermodal freight traffic surged toward a new record.

RSIA mandated that FRA, as the Secretary's designee, complete an unprecedented 42 tasks, including final rules, guidance documents, model State laws, studies, and reports as well three types of annual reports and hundreds of periodic accident reporting audits.

Thirty of the 42 tasks are complete, and the rest are in the pipeline progressing towards completion. Appendix 1 lists the rulemakings, non-periodic reports and studies, guidance, and model State laws that FRA has completed as of February 26, 2014.

The chart and table below illustrate a decade of safety improvement.

Response to Accidents

As we use data and research to drive continuous safety improvement, we learn from every accident. FRA investigators focus on identifying an accident's root causes so we can further eliminate risk and take appropriate enforcement action. This is one more facet of our comprehensive approach to rail safety.

The Department, including FRA, has responded aggressively to recent accidents that have received widespread attention.

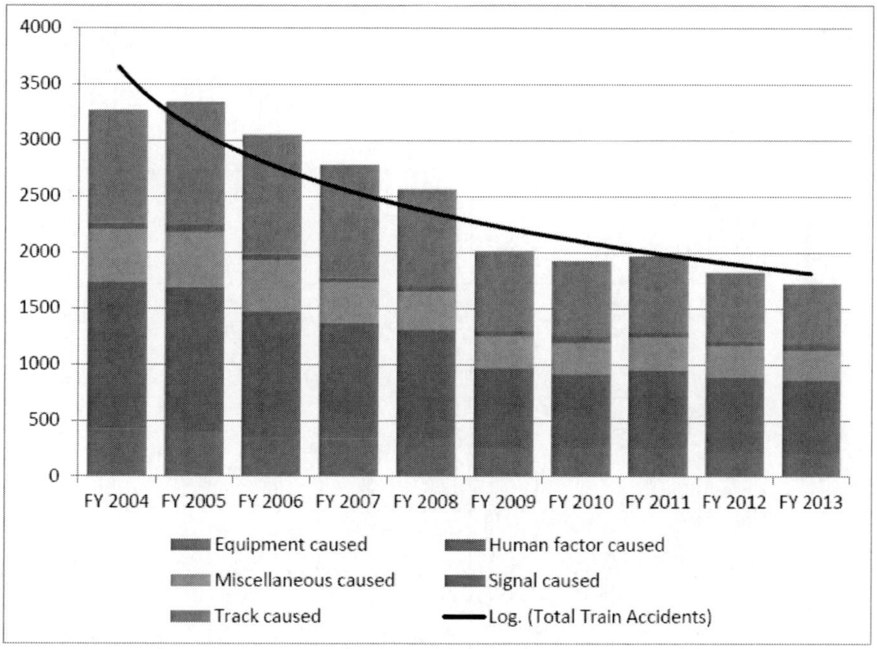

* Fiscal Year Representing Absolute Numbers.

Ten-year Downward Trend for Train Accident Reductions (FY04-FY13).

Ten-year Railroad Safety Trends by Accident/Incident Cause

	FY 2004	FY 2005	FY 2006	FY 2007	FY 2008	FY 2009	FY 2010	FY 2011	FY 2012	FY 2013
Total Accidents/ Incidents	19.039	18.093	17.525	17.298	16.907	16.873	16.696	16.063	15.167	14.852
Human-Factor-Caused Train Accidents	1.721	1.648	1.380	1.297	1.230	1.041	0.948	0.995	0.919	0.888
Track-Caused Train Accidents	1.314	1.398	1.318	1.258	1.094	1.036	0.972	0.954	0.843	0.727
Equipment-Caused Train Accidents	0.548	0.499	0.433	0.418	0.435	0.366	0.370	0.342	0.286	0.271
Total Signal/ Misc.- Caused Train Accidents	0.692	0.707	0.641	0.506	0.497	0.484	0.494	0.469	0.438	0.430
Highway-Rail Incidents	4.024	3.800	3.797	3.523	3.240	2.986	2.900	2.881	2.773	2.685
Non-Accident Hazmat Releases	1.387	1.398	1.147	1.221	1.227	1.149	1.063	1.079	0.933	0.932

* Accident/Incident, Train Accident, and Highway-Rail Incident Numbers Normalized by Million Train-Miles for Fiscal Year, Non-Accident Hazmat Releases Normalized by 200 Million Hazmat Ton-Miles for Fiscal Year.

***Metro-North Commuter Railroad Company*[1]**

As a result of several accidents on Metro-North Commuter Railroad Company (Metro-North), FRA issued Emergency Order 29 and Safety Advisory 2013-08 on December 11, 2013.

- Emergency Order 29 required Metro-North to take immediate action to prevent excessive train speeds by identifying and prioritizing high-risk areas, modifying its existing signal system to ensure speed limits are obeyed, and ensuring a higher level of engagement and communication among operating crewmembers in higher risk locations. To date, FRA has not identified any instances of noncompliance with Emergency Order 29.
- Safety Advisory 2013-08 helps ensure that all railroads adhere to Federal regulations regarding maximum authorized train speed limits through training, operational testing, and train crewmember communication.

On December 16, 15 days after a fatal accident in New York, FRA commenced Operation Deep Dive, a comprehensive, multi-disciplinary safety assessment of Metro-North where technical and human factors experts are reviewing safety-critical procedures and processes, including operations, mechanical and engineering. The Federal Transit Administration is participating with FRA to ensure investments in Metro-North are properly prioritized to improve safety.

The rail safety team is assessing the following:

- Track, signal and rolling stock maintenance, inspection and repair practices;
- Protection for employees working on rail infrastructure, locomotives and rail cars;
- Communication between mechanical and transportation departments at maintenance facilities;
- Operation control center procedures and rail traffic controller training;
- Compliance with Federal hours of service regulations, including fatigue management programs;
- Evaluating results of operational data to measure efficiency of employees' execution and comprehension of all applicable Federal rail safety regulations;

- Locomotive engineer oversight;
- Engineer and conductor certification; and
- Operating crew medical requirements.

Operation Deep Dive ended February 14, 2014 and FRA will present a report of its findings within 30 days afterwards. FRA will meet with Metro-North to discuss the findings and appropriate remedial actions. Additionally, FRA will discuss best practices and lessons learned from Operation Deep Dive with other commuter rail chief executive officers (CEOs) through the American Public Transportation Association.

Rail Accidents involving Crude Oil

Crude oil transportation by rail rose quickly because of increasing production in the Bakken region of North Dakota. FRA is paying close attention to that region, and accident rates in North Dakota have fallen over the past three years, even with increased traffic.

In response to recent train accidents in the United States and Canada involving tank cars carrying crude oil, DOT, including FRA and the Pipeline and Hazardous Materials Safety Administration (PHMSA), has taken action on multiple fronts to mitigate risks and ensure the safe transportation of crude oil, ethanol, and other hazardous materials by rail. FRA and PHMSA have related but distinct responsibilities in managing the risk from the transportation of hazardous materials. PHMSA produces regulations pertaining to the transportation of hazardous materials by rail, which are primarily enforced by FRA's safety staff, while FRA's staff also acts to enforce comprehensive safety regulations for rail transportation.

On January 16th, oil industry representatives and rail industry CEOs met with the Secretary and heads of PHMSA, the Federal Motor Carrier Safety Administration, and FRA in a "Call to Action." The CEOs were asked to develop specific plans to immediately improve the safety of crude oil shipments, and recommendations on how to improve safety over the long term. After analyzing their plans and suggestions, on February 20, 2014, Secretary Foxx sent a letter to the Association of American Railroads (AAR) with a list of actions to be voluntarily taken immediately by industry to dramatically improve the safety of railroads transporting crude oil and the communities they move through. AAR President and CEO Edward Hamberger signed the agreement that same day, and individual railroads are signing on subsequently. The letter from Secretary Foxx listed eight commitments:

1. By July 1, subscribers will apply HAZMAT routing analysis to trains with 20 or more tank cars loaded with petroleum crude oil (Key Crude Oil Trains). The routing analysis utilizes a computer model to analyze 27 risk factors to determine the safest and most secure route for the product to travel.
2. By July 1, subscribers will adhere to a speed restriction of 50 mph for all Key Crude Oil Trains, and 40mph in high-threat urban areas if they are using a DOT 111 tank car.
3. By April 1, subscribers will equip all Key Crude Oil Trains on main track with distributive power locomotives or an operative two-way telemetry end of train device to achieve benefits in braking speed and substantially reducing the kinetic energy in trains to prevent pile ups.
4. Effective March 25, subscribers will perform at least one internal rail inspection and two track geometry inspections more than is required by current regulations every calendar year on Key Crude Oil Train routes.
5. By July 1, subscribers will begin installing wayside defective bearing detectors every 40 miles on Key Crude Oil Train routes to prevent equipment-caused accidents.
6. Subscribers will develop an inventory of emergency response resources along Key Crude Oil Train routes. This information will be provided to DOT and emergency responders upon request.
7. Subscribers will provide $5 million to develop and provide training on hazardous material transportation and fund training for emergency responders through the end of 2014. Comprehensive training will occur at the Transportation Technology Center, Inc. facility in Colorado with a training program fully developed by July 1.
8. Subscribers will continue to work with communities on Key Crude Oil Train routes to address location-specific concerns.

A copy of the full agreement is included with this testimony. This agreement is an important step in improving the safety of crude oil transportation by rail. FRA will continue to use its regulatory authority to address this issue and act accordingly to maintain public safety and confidence.

Here is a summary of other DOT actions in response to accidents involving crude oil and other hazardous materials.

Order and Advisories

FRA issued Emergency Order 28, and both FRA and PHMSA issued safety advisories, held public hearings, and notified shippers and carriers of the critical importance of public safety when transporting hazardous materials.

- FRA's emergency order addresses unattended trains, train securement, the use of locks, communication between train crews and dispatchers, and daily safety briefings for railroad employees and was published August 7, 2013.
- A joint FRA-PHMSA safety advisory on related issues was also published August 7, 2013.
- A joint FRA-PHMSA follow-up safety advisory was published November 20, 2013.

Rulemakings

In addition to the emergency order and safety advisories, FRA is updating applicable rail safety regulations, and as PHMSA will describe in more detail, FRA is collaborating with PHMSA on a rulemaking that addresses DOT Specification 111 tank cars. All rulemakings are subject to extensive study and analysis.

But tank cars are only one part of the chain of delivery, and we must identify and evaluate *all* of the risks associated with bulk movements of hazardous material, such as ethanol and crude oil, and then work to eliminate those risks.

- On August 28, 2013, FRA and PHMSA held a public meeting with industry stakeholders to solicit input for a comprehensive review of the Hazardous Materials Regulations applicable to rail. PHMSA and FRA are collaborating to address comments received at the public meeting.
- On August 29, 2013, FRA convened an emergency session of the RSAC. During the emergency RSAC meeting, participants established three collaborative working groups to formulate new rulemaking recommendations regarding (1) transportation of hazardous materials by rail, (2) appropriate train crew sizes, and (3) train securement procedures. These working groups are meeting on a regular basis and

we expect formal recommendations for consideration by April 1, 2014.

Operation Classification (The "Bakken Blitz")

In August 2013, PHMSA, supported by FRA, launched Operation Classification, which involves joint activities at all transportation phases to investigate how shippers and carriers are classifying crude oil and what actions they are taking to understand the characteristics of the material. The operations have primarily targeted shipments from the Bakken region and consisted of unannounced spot inspections, data collection, and sampling as well as verifying compliance with Federal safety regulations. Operation Classification is nearing completion.

As I have described, rail safety is at an all-time best. Yet, these accidents illustrate why we can never be complacent.

Our Vision for the Next Generation of Rail Safety

Continuous safety improvement requires a comprehensive strategy designed to eliminate risk. Here is FRA's strategy, founded on three pillars:

1. Continuing a rigorous regulatory and inspection program based on strategic use of data;
2. Advancing proactive approaches for early identification and reduction of risk; and
3. Capital investments, and robust research and development.

PILLAR I. CONTINUING A RIGOROUS REGULATORY AND INSPECTION PROGRAM

As stated previously, FRA's approach to rail safety has led to unprecedented safety improvements. We will continue this framework for safety oversight and enforcement and improve it. Data driven analysis will continue to guide workforce planning and inspection activities.

FRA's regulatory program improves safety by developing rules based on facts, incident and accident causation analysis, comparison of alternative mitigation measures, and cost-beneficial solutions. FRA rulemaking considers current and future industry capabilities, compliance burden and cost, and other economic and social realities. Within this context, FRA will continue to attempt to meet statutory milestones with its available resources.

State rail inspectors are a force multiplier for FRA's compliance and enforcement efforts. The State Rail Safety Participation Program consists of States employing safety inspectors in the five rail safety inspection disciplines. State programs conduct planned, routine compliance inspections; and may undertake additional investigative and surveillance activities consistent with overall program needs and individual State capabilities. FRA provides on-the-job training to State inspectors. We invite additional state participation in this important program and view it as an opportunity to improve oversight in key states and regions.

Focus Areas

Safety overall has improved; however, accidents related to human error and track defects account for more than two-thirds of all train accidents, and trespassing and highway-rail grade crossing incidents account for approximately 95 percent of all rail-related fatalities. We will allocate resources and work with partners, such as Operation Lifesaver, to make improvements in these challenging areas. The following rulemakings, reports, guidance documents, and other actions are important milestones that will guide our work in these areas:

Human Factors

- Final rule to advance nationwide implementation of positive train control (PTC) systems (which prevent overspeed derailments, train-to-train collisions, and other types of accidents often caused by human error) by defining statutory terms and the essential functionalities of PTC systems. FRA also issued two other rules designed to reduce some of the costs of PTC implementation,. PTC systems are a technology that promotes safety improvement through the reduction of certain human-factor-related incidents and will complement FRA's other safety efforts, such as implementation of safety Risk Reduction Programs (RRP) and crash energy management.

- Final rule requiring a railroad to have a formal program for certifying train conductors. This will raise the bar of professionalism and ensure that only those persons who meet minimum Federal safety standards serve as conductors.
- Proposed rule that would enhance safety by mandating that certain railroads (each Class I railroad, intercity passenger railroad, and commuter railroad) have a Critical Incident Stress Plan that may help mitigate the long-term negative effects of critical incidents upon railroad employees and the impact of performing safety-sensitive duties in the days following such incidents when the associated stress may hinder their ability to perform such duties safely.
- Final rule on the hours of service of passenger train employees. This rule draws on detailed research into the causes of train operator fatigue and analysis of thousands of operator work patterns. FRA also published in the *Federal Register* three lengthy, detailed statements of agency policy and interpretation to clarify the hours of service laws as amended by RSIA.
- An FRA-led industry-wide initiative to combat the dangers of electronic device distraction in the railroad workplace as well as an emergency order and then a final rule prohibiting distracted operation of trains.
- A proposed rule that would establish minimum training standards for each class or craft of safety-related employee and contractor. The rule would require the qualification and documentation of the proficiency of such employees on their knowledge and ability to comply with Federal railroad safety laws and regulations and the employing railroad company's rules and procedures implementing those laws and regulations. A final rule on minimum training standards and plans is under development.

Track Safety

- Final rule to Improve Rail Inspections. Requires the use of performance-based rail inspection methods that focus on maintaining low rail failure rates per mile of track and generally results in more frequent testing; provides a four-hour period to verify that certain less serious suspected defects exist in a rail section once track owners learn that the rail contains an indication of those defects; requires that rail inspectors are properly qualified to operate rail flaw detection equipment and interpret test results; and establishes an annual maximum allowable rate of rail defects and rail failures between inspections for each designated inspection segment of track. These changes are intended to reduce the risk of derailments caused by rail failures by improving the accuracy of rail inspections and shortening the time that latent, undetected rail flaws remain in track.

- Vehicle/Track Interaction Safety Standards. The final rule was based on research into vehicle/track interaction, and it promotes the safe interaction of rail vehicles with the track over which they operate under a variety of conditions at speeds up to 220 mph. The rule also adds flexibility for safely permitting high cant deficiency train operations[2] through curves at more conventional speeds so that both freight and passenger trains may better sustain maximum allowable speeds through curved track.
- New Technology to Improve Track Safety. Through our research and development program we are about to bring to market new technology for avoiding track buckles (sun-kinks). The device measures the neutral temperature of rail and warns the railroad when track maintenance is required to avoid track buckling. We are also developing technology to predict rail temperature variations. This provides railroads information needed to decide the extent and duration of slow orders to reduce safety risk on hot days.

Grade Crossing Safety and Trespass Prevention

- Standards requiring railroads to establish and maintain toll-free "1-800" emergency notification systems by which the public can telephone the proper railroad about a stalled vehicle or other safety problem at a specifically identified grade crossing.
- Regulations requiring 10 States to issue State-specific action plans to improve safety at highway- rail grade crossings.
- Model State laws on highway users' sight distance at passively signed crossings and on highway motorists' violations of grade crossing warning devices.
- A proposed rule specifying the types of information that railroads would have to report to the Department's National Crossing Inventory.

- A five-year strategy to improve highway-rail grade crossing safety, including an audit every two years of Class I railroads' highway-rail grade crossing accident reports to ensure that these railroads are accurately reporting these incidents. Resources permitting, FRA will conduct such audits every five years on other railroads.
- Guidance addressing pedestrian safety at or near passenger rail stations,
- An FRA-released smartphone application with grade crossing information.

PILLAR II. ADVANCING PROACTIVE APPROACHES TO REDUCE RISK

Continuous safety improvement requires a multi-faceted approach. The next level of safety will come from advancing proactive safety-based programs that analyze risks, identify hazards, and put in place customized plans to eliminate those risks.

- Risk Reduction Programs (RRP) and System Safety Programs (SSP) that help identify accident precursors so that corrective action can be taken in advance. We will issue a final rule before the end of 2014 to require passenger railroads to develop and implement SSPs. A notice of proposed rulemaking that would require freight railroads to establish RRPs is currently under development. Both are designed to require railroads to develop and implement systematic risk- based approaches to ensuring continuous safety improvement.
- Confidential Close Call Reporting System (C^3RS), a voluntary and non-punitive program for railroads and their employees to report close calls. Results from one C^3RS pilot site indicate nearly a 70-percent reduction in certain accidents. C^3RS helps develop a positive and proactive safety culture, using detailed data far beyond what is obtained during accident investigations. The magnitude of the information provided from proactive programs like C^3RS in comparison to traditional data from accidents and injuries is illustrated below:

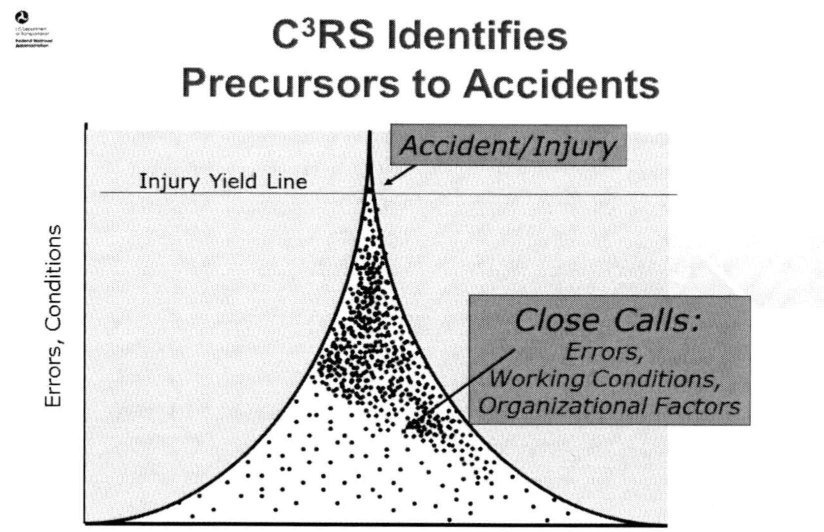

Programs like Confidential Close Calls Reporting allow us to gather data *before* an accident occurs and to develop risk mitigation strategies well in advance.

PILLAR III. CAPITAL INVESTMENTS, INCLUDING ROBUST RESEARCH AND DEVELOPMENT

As you know, portions of two important rail laws expired at the end of FY 2013: RSIA and the Passenger Rail Investment and Improvement Act of 2008 (PRIIA). The President's FY 2014 budget for FRA laid out a comprehensive, multi-year reauthorization blueprint for moving forward. The fundamental goal of this proposal is to develop a coordinated approach to enhancing the Nation's rail system–an integrated strategy that addresses safety and passenger and freight service improvements. This new approach reflects the complex reality of how rail works in the United States–most track is privately-owned and carries a mix of passenger and freight trains. Safety is improved not just through regulations and inspections but also through capital investments and research and development.

For example, chokepoints often hinder the efficient movement of intercity passenger, commuter, and freight trains, while the elimination of grade crossings with strategic placement of overpasses and underpasses enhances rail, vehicular, and pedestrian safety.

FRA's reauthorization proposal's key priorities include the following:

- **Modernizing our rail infrastructure**. Past generations of Americans invested heavily in building the infrastructure we rely on today. Most segments of the Northeast Corridor were built more than a century ago. Maintaining and modernizing these assets will lower long-term costs and result in a safer, more efficient and reliable rail system.
- **Meeting the growing market demand**. With 100 million more Americans expected by 2050, the national transportation system must be prepared to handle substantial increases in the movement of people and goods. Given the existing capacity constraints on other modes, rail will play an increasingly vital role in balancing America's transportation system by accommodating this growth, resulting in public benefits such as reduced reliance on foreign oil, reduced air pollution, increased safety, and more travel options. The budget incorporates market-based investments in building or improving passenger rail corridors, eliminating rail chokepoints, adding freight capacity, and conducting comprehensive planning.
- **Successfully implementing PTC.** The mandated deadline of December 2015 will likely not be reached by many railroads. Commuter rail operations are cash-strapped and unable to attain certain necessities for implementation, such as communications spectrum. FRA's budget proposes grants for those commuter railroads and research and development for new technologies to improve rail safety. FRA's August 2012 Report to Congress "Positive Train Control: Implementation Status, Issues, and Impacts" summarized the major technical and programmatic challenges and obstacles associated with PTC implementation that FRA had identified so far.[3] Subsequent to the report's submission, a new issue regarding PTC communications towers deployment arose under the jurisdiction of the Federal Communications Commission.
- **Promoting innovation.** FRA's vision is for the domestic rail industry to be again world-leading. We want U.S. companies to develop patents for state-of-the-art rail technology, to supply rail operators throughout the world, and to employ the best engineers and railway workers. The United States should be exporting intellectual capital and rail products, not importing them.

- **Mitigating rail's impacts on communities.** Improving quality of life by eliminating grade crossings, sealing corridors, reducing noise impacts, and including safety enhancements that allow for service improvements and economic growth.
- **Research and Development.** Implementing new technology will be a key driver for future safety improvement. Here are a few examples of important research:
  - Track inspection technologies that detect defects before they become failures in service.
  - Computer modeling capabilities to improve understanding of vehicle/track interaction, wheel and rail profiles, and contact conditions.
  - Autonomous recording methods to provide more frequent and cost-effective measurements of track condition.
  - Research to develop new methods for monitoring difficult-to-detect safety issues such as longitudinal rail force, ballast lateral restraint, and ballast condition.
  - High-speed rail research and development, which has identified several key risk factors for corridors shared by passenger and freight operations. Research to understand these risks and mitigate them is ongoing.
  - Research on new technologies for improving grade crossing safety. One project that has significant potential is implementation of Intelligent Transportation Systems at grade crossings. FRA is also conducting human-factors research to understand the behavior of highway users when they approach grade crossings. This research is expected to lead to recommendations for improved signage and warning systems. FRA will consider the benefits and costs, and feasible alternatives, for any recommendation.
  - A research and development program to achieve reliable, long life from concrete ties. The program involves freight railroads, Amtrak, manufacturers, and universities.
  - The National Cooperative Rail Research Program, which enhances the development of technical skills for a capable workforce to design and operate the next generation of safe railroads.

The Need for Predictable Funding

An overarching issue that runs across all of these priorities is the need for sustained and predictable Federal funding for rail programs, similar to the treatment of other modes of transportation. Congress has for decades funded highway infrastructure and safety, transit, and aviation programs through multi- year authorizations that provide guaranteed funding. This enables States, local governments, and other stakeholders to plan and make large-scale infrastructure investments on a year-to-year basis. Likewise, internationally, other major rail systems have been planned and developed through a predictable multi- year funding program.

CONCLUSION

Thank you for the opportunity to testify and answer your questions today. Safety is FRA's number one priority, and we appreciate your attention and focus on such an important issue for the American public. Our vision for the next generation of rail safety balances a comprehensive and effective regulatory framework with innovative, proactive ideas and capital investment, including critical research and development. We look forward to working with this Committee to improve our programs and make the American rail network as safe, reliable, and efficient as possible. I will be happy to respond to your questions.

APPENDIX 1.

FRA Rulemakings Completed as of March 5, 2014, that Were Mandated, Explicitly or Implicitly, by RSIA[4]

1. To specify the essential functionalities of mandated PTC systems, define related statutory terms, and identify additional lines for implementation. *(Sec. 104)*.[5]
2. To establish substantive hours of service requirements for passenger train employees. *(Sec. 108(d))*.

3. To update existing hours of service recordkeeping regulations. *(Sec.108(f))*.
4. To require State-specific action plans from certain States to improve safety at highway-rail grade crossings. *(Sec. 202)*.
5. To require toll-free telephone emergency notification numbers for reporting problems at public and private highway-rail grade crossings. *(Sec. 205)*.
6. Increase the ordinary maximum and aggravated maximum civil penalties per violation for rail safety violations to $25,000 and $100,000, respectively. *(Sec. 302)*.
7. On prohibition of individuals from performing safety-sensitive functions in the railroad industry for a violation of hazardous materials transportation law. *(Sec. 305)*.
8. On procedures for emergency waivers. *(Sec. 308)*.
9. To require the certification of conductors. *(Sec. 402)*.
10. On the results of FRA's study of track inspection intervals and other track issues. *(Sec. 403(c))*.
11. On concrete ties. *(Sec. 403(d))*.
12. To require owners of railroad bridges to implement programs for inspection, maintenance, and management of those structures. *(Sec. 417)*.
13. On camp cars used as railroad employee sleeping quarters. *(Sec. 420)*.
14. Amending regulations of the Office of the Secretary of Transportation to provide that the Secretary delegates to the Administrator of FRA the responsibility to carry out the Secretary's responsibilities under RSIA.

Completed RSIA-Mandated Guidance and Model State Laws[6]

1. Guidance on pedestrian safety at or near rail passenger stations. *(Sec. 201)*.
2. Guidance for the administration of the authority to buy items of nominal value and distribute them to the public as part of a crossing safety or railroad trespass prevention program. *(Sec. 208(c))*.
3. Model State law on highway users' sight distances at passively signed highway-rail grade crossings. *(Sec. 203)*.

4. Model State law on motorists' violations of grade crossing warning devices. *(Sec. 208).*

Completed RSIA-Mandated Non-periodic Reports or Studies

1. Report to Congress on DOT's long-term (minimum 5-year) strategy for improving rail safety, including annual plans and schedules for achieving specified statutory goals, to be submitted with the President's annual budget. *(Sec. 102).*
2. Report to Congress on the progress of railroads' implementation of PTC. *(Sec. 104).*
3. Conduct study to evaluate whether it is in the public interest to withhold from discovery or admission, in certain judicial proceedings for damages, the reports and data compiled to implement, etc., a required risk reduction program. *(Sec. 109*).
4. Evaluate and review current local, State, and Federal laws regarding trespassing on railroad property, vandalism affecting railroad safety, and violations of highway-rail grade crossing warning devices. *(Sec. 208(a)).*
5. Report to Congress on the results of DOT research about track inspection intervals, etc. *(Sec. 403(a)- (b)).*
6. Conduct study of methods to improve or correct passenger station platform gaps *(Sec. 404*).
7. Report to Congress detailing the results of DOT research about use of personal electronic devices in the locomotive cab by safety-related railroad employees. *(Sec. 405).*
8. Report to Congress on DOT research about the effects of repealing a provision exempting Consolidated Rail Corporation, etc., from certain labor-related laws (45 U.S.C. § 797j). *(Sec. 408).*
9. Report to Congress on the results of DOT research about exposure of railroad employees and others to radiation. *(Sec. 411*).
10. Report to Congress on DOT study on the expected safety effects of reducing inspection frequency of diesel-electric locomotives in limited service by railroad museums. *(Sec. 415).*
11. Report to Congress on model plans and recommendations, to be developed through a task force to be established by DOT, to help railroads respond to passenger rail accidents. *(Sec. 503).*

APPENDIX 2.

Metro-North Commuter Railroad Company (Metro-North) is the second largest commuter railroad in the nation, with an annual ridership of 82,953,628.[7] It is a subsidiary agency of the Metropolitan Transportation Authority, a New York State Authority.

- Three main lines, the Hudson, Harlem, and New Haven Lines, branch northward out of Grand Central Terminal, located in mid-town Manhattan, into suburban New York and Connecticut. Metro-North maintains the equipment and infrastructure and operates and controls the trains on these lines.

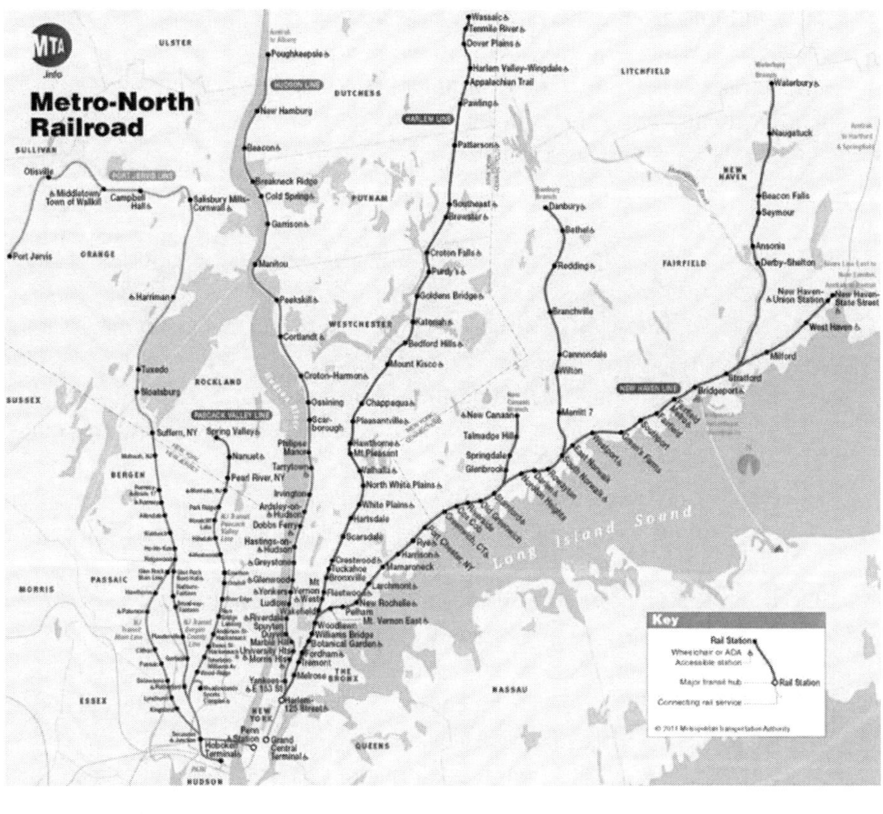

Map of the Metro-North System.

- Amtrak operates on the Hudson Line, between Spuyten Duyvil and Poughkeepsie, and on the New Haven Line, between New Rochelle and New Haven.
- The West of Hudson Service, the Port Jervis and the Pascack Valley Lines, operates from New Jersey Transit Rail Operations' (NJ Transit) Hoboken terminal, providing service to Rockland and Orange counties. NJ Transit maintains the equipment and operates and controls the trains. Metro-North maintains the infrastructure.

End Notes

[1] *A description of Metro-North Railroad is in Appendix 2 to this testimony.*

[2] Cant deficiency involves traveling through a curve faster than the balance speed and produces a net lateral force to the outside of the curve. http://www.highspeed- rail.org/Documents/ PRIIA%20305%20DocSpec%20and%20other%20NGEC%20Documents/305%20PRIIA% 20Tilt%20pre sentation.pdf

[3] "Positive Train Control: Implementation Status, Issues, and Impacts" - http://www.fra.dot.gov/ Elib/Details/L03718

[4] In addition, FRA commenced a rulemaking to define "critical incident" for purposes of the mandated rulemaking on critical incident stress plans as specifically required by Sec. 410(c)).

[5] In addition, FRA has issued two final rules on PTC, and another final rule on PTC is in clearance in the Executive Branch.

[6] In addition, FRA has published three guidance documents on the hours of service laws as amended by RSIA in the *Federal Register*.

[7] http://web.mta.info/mta/network.htm#statsmnr

In: Transporting Crude Oil by U.S. Rail
Editor: Rosario S. McLaughlin ISBN: 978-1-63117-837-5

Chapter 3

STATEMENT OF CYNTHIA L. QUARTERMAN, ADMINISTRATOR, PIPELINE AND HAZARDOUS MATERIALS SAFETY ADMINISTRATION. HEARING ON "ENHANCING OUR RAIL SAFETY: CURRENT CHALLENGES FOR PASSENGER AND FREIGHT RAIL"*

Good morning Mr. Chairman, Ranking Member, and Members of the Subcommittee. Thank you for the opportunity to appear today to discuss the Pipeline and Hazardous Materials Safety Administration's (PHMSA) comprehensive approach to address the risks associated with increased bulk shipments of flammable liquids by rail. I would also like to thank you for your leadership and for your efforts to advance rail safety. While rail safety is improving, high-profile train accidents like the ones we've seen in Lac-Mégantic, Quebec, Canada; Aliceville, Alabama; and Casselton, North Dakota underscore how important it is to be ever-vigilant in protecting local communities and the environment.

Safety is the top priority for Secretary Foxx, everyone at PHMSA, and the other modes in the U.S. Department of Transportation (DOT). PHMSA continues to work diligently to protect the American people and the

* This is an edited, reformatted and augmented version of a statement presented March 6, 2014 before the Senate Committee on Commerce, Science, and Transportation, Subcommittee on Surface Transportation and Merchant Marine Infrastructure, Safety, and Security.

environment from the risks of hazardous materials transportation by all modes, including rail. PHMSA works to achieve its safety mission through efforts to prevent and mitigate accidents by developing regulations and guidance, taking rigorous enforcement actions, collaborating with stakeholders, and educating emergency responders and the public.

This testimony will focus on the risks posed by the transport of bulk shipment of flammable liquids, including petroleum crude oil, by rail and PHMSA's efforts to both prevent and mitigate those risks. First, I will provide an overview of the current state of petroleum crude oil (crude oil) transportation in the United States. Second, I will discuss our comprehensive approach to prevent and mitigate the damage caused by rail accidents involving hazardous materials.

I. State of Crude Oil Transportation by Rail

As energy production in the United States increases, so does the transportation of more products in their various forms by multiple modes. The epicenter of the increased crude oil production is the Bakken Formation, occupying about 200,000 square miles (520,000 square kilometers^2) of the subsurface underlying parts of Montana and North Dakota, and Saskatchewan and Manitoba in Canada. Production from the Bakken in recent years has elevated North Dakota to the second largest oil producing State, and it is one of the most important sources of oil in the United States. While most new Bakken drilling and production has been in North Dakota, drilling operations also extend into Montana, Saskatchewan, and Manitoba. As of 2013, the Bakken produced more than ten percent of all oil in the United States. In November 2013, 10,022 Bakken wells extracted approximately 29 million barrels of oil and 32 million cubic feet of gas. This equates to over 900,000 barrels of oil produced daily (See Table 1).[1]

Approximately 71 percent of all oil produced in North Dakota, or around 800,000 barrels per day, was transported by rail in November 2013, according to the North Dakota Public Service Commission.[2] Corresponding with increased production, the volume of crude oil moving by rail has quadrupled in less than a decade (See Table 2—a Class I railroad is a railroad having annual inflation-adjusted operating revenues for three consecutive years of

$250 million or more as the figure $250 million is adjusted by applying the railroad revenue deflator formula).[3] While overall train volume has increased, train accidents declined by 43 percent, and train accidents involving a hazardous materials release are down 16 percent between 2003 and 2012.[4]

Despite this decline in such accidents, there is always the potential for low-probability, high-consequence events to occur that could have devastating consequences to the public, communities, and the environment. Recent incidents in the United States and Canada demonstrate the need for a renewed focus on rail safety efforts. PHMSA works in partnership with the Federal Railroad Administration (FRA) to address and mitigate the risks associated with the rail transport of hazardous materials. Here are a few incidents that have sharpened our focus on the safe transportation of high-hazard flammable liquids by rail.

On July 6, 2013, a catastrophic derailment involving an unattended freight train containing 72 loaded DOT Specification 111 tank cars of crude oil occurred in the town of Lac- Mégantic, Quebec. The train, which was 4,701 feet long and weighed 10,287 tons, was not properly secured and rolled down a descending grade, subsequently derailing near the center of Lac-Mégantic. The locomotives separated from the train and came to a stop about a half mile east of the derailment. In the course of the accident, 63 tank cars derailed. Several derailed tank cars released crude oil, causing fires that killed 47 people, extensively damaged the town center, and required the evacuation of about 2,000 people from the surrounding area. Transport Canada is currently investigating the accident, with the assistance of the National Transportation Safety Board (NTSB) and DOT.

On November 7, 2013, a train carrying crude oil to the Gulf Coast from North Dakota derailed in Aliceville, Alabama, spilling crude oil in a nearby wetland and igniting into flames. There were a total of 88 DOT 111 tank cars containing crude oil in the 90-car train. Twenty-six DOT Specification 111 tank cars derailed, 21 of which released all or part of their contents. The NTSB is currently investigating the accident, with the assistance of DOT.

On December 30, 2013, a train carrying crude oil derailed and ignited near Casselton, North Dakota, prompting authorities to issue a voluntary evacuation of the city and surrounding area. A collision with a disabled train blocking the track caused 20 DOT Specification 111 tank cars to derail. Estimates indicate that those cars lost 476,436 gallons of product. NTSB is investigating the accident, with the assistance of DOT.

Table 1. Bakken Oil Production in North Dakota 2003-2013

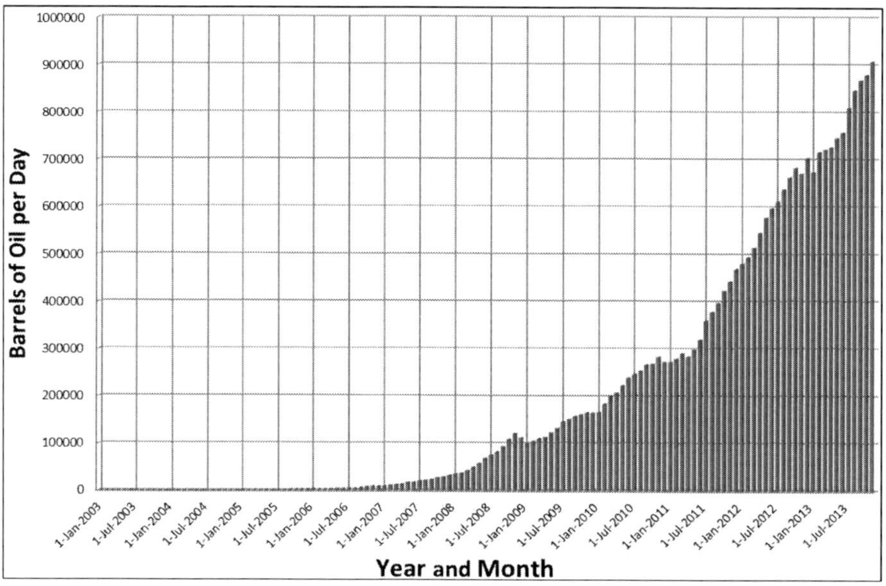

Table 2. Originated Carloads of Crude Oil on Class I Railroads

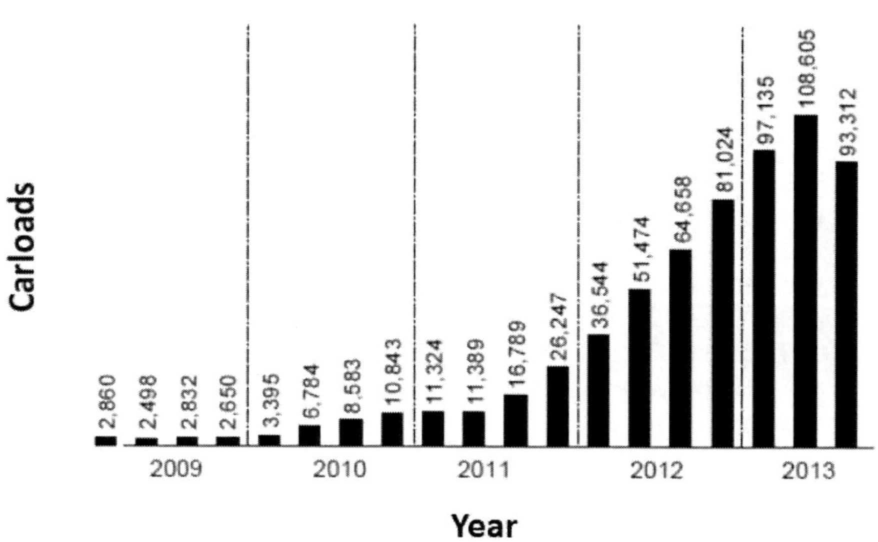

Accidents like these demonstrate both the inherent dangers of transporting hazardous materials and the various factors that may cause accidents and the unintentional release of hazardous materials. PHMSA strives to prevent these accidents from occurring and, in the event they do occur, helps to mitigate the consequences of these types of accidents. Train accidents involving hazardous materials releases like the ones previously described highlight the need for a robust hazardous materials transportation regulatory system, strong enforcement capabilities, and wide-reaching communication with and training of hazardous materials stakeholders, including the public, hazardous materials transporters, and emergency responders.

II. Comprehensive Approach to Prevent and Mitigate Rail Hazardous Materials Accidents and Incidents

PHMSA's safety mission involves working to ensure that the transportation system is functioning as it should. With regard to rail safety, PHMSA and FRA have taken a comprehensive approach to mitigating the risks posed by the bulk transport of hazardous materials by rail. Specifically, PHMSA, in coordination with FRA, is focusing on methods to prevent accidents and incidents from occurring and ways to mitigate the effects of those events that do occur. On the prevention front, we are working together to implement necessary operational controls and ensure rail track integrity to lessen the likelihood of accidents. PHMSA has requirements in place to mitigate effects of potential accidents through appropriate classification of the materials being transported; appropriate packaging of the materials, including ensuring materials are in the appropriate container; and effectively communicating to transportation workers and first responders what material is involved so they can handle or respond correctly to any accidents associated with the material. This approach is designed to prevent the occurrence of a hazardous materials release in the course of rail transportation and mitigate the damage caused should a hazardous material release occur.

PHMSA has a variety of regulatory and non-regulatory tools to address the risks of the bulk transport of flammable materials, including crude oil, by rail. In the wake of increased crude oil movements by rail and recent incidents,

PHMSA has used many of these tools to improve safety. Most recently, PHMSA has issued guidance and an advanced notice of proposed rulemaking, participated in rail safety committees, held public meetings, enhanced enforcement and inspection efforts, and coordinated with other agencies to improve the safety of the public.

Regulatory Efforts by PHMSA and FRA

On May 14, 2010, PHMSA published a final rule (HM-233A) to amend the Hazardous Materials Regulations to incorporate provisions contained in certain widely used or longstanding special permits that have an established safety record. [5] As part of that rulemaking, PHMSA adopted a requirement that would allow certain rail tank cars transporting hazardous materials to exceed the gross weight on rail limitation of 263,000 pounds upon approval by FRA.

On January 25, 2011, FRA issued a *Federal Register* notice of FRA's approval pursuant to PHMSA's May 14, 2010 final rule.[6] The approval established detailed conditions for manufacturing and operating certain tank cars in hazardous materials service, including the DOT Specification 111 tank car, which is the tank car used for the transportation of flammable liquids, such as crude oil, that weigh between 263,000 and 286,000 pounds. These actions provided tank car manufacturers with the authority to build a 286,000-pound tank car. Rail car manufacturers have used that authority to manufacture an enhanced DOT Specification 111 tank car (CPC-1232) under the conditions outlined in the January 25, 2011 approval. Specific improvements to the car include the following: normalized steel, puncture resistance, head shields, and top fitting protection. Should a manufacturer choose to design a car outside the conditions of that approval, it can seek another approval in accordance with section 179.13 of the Hazardous Materials Regulations in title 49 of the Code of Federal Regulations. To date, PHMSA and FRA have not received any requests to design a car that deviates from the January 25, 2011 approved design.

Following the publication of the PHMSA final rule and the subsequent FRA approval, PHMSA received a petition (P-1577)[7] from the Association of American Railroads (AAR) on March 9, 2011, requesting changes to PHMSA's specifications for the DOT Specification 111 tank car used to transport Packing Group I and II materials[8] (See Table 4 for tank car comparison).[9] During the summer of 2011, the AAR Tank Car Committee (TCC) created a task force (Task Force), which included PHMSA and FRA participation, with a dual responsibility to develop an industry standard for tank cars used to transport crude oil, denatured alcohol, and ethanol/gasoline mixtures and to consider operating requirements to reduce the risk of derailment of tank cars carrying crude oil classified as Packing Group I and II, or ethanol.

The Task Force worked to address the root cause, severity, and consequences of derailments, and its recommendations were finalized on March 1, 2012. As a result PHMSA, with FRA's agreement, initiated an advance notice of proposed rulemaking (ANPRM) to arrive at a more comprehensive solution.

In May 2012, PHMSA began drafting an ANPRM to consider revisions to the Hazardous Materials Regulations to improve the crashworthiness of railroad tank cars and identify and address operational improvements. The draft ANPRM addressed several Petitions for Rulemaking submitted by industry and recommendations issued by the NTSB.[10] The draft ANPRM posed a series of questions to the regulated community designed to solicit comments on potential operational and tank car design improvements that could improve rail safety, along with the costs of these improvements. The draft ANPRM was also designed to build and improve on the Task Force recommendations and examined the differences in the DOT-approved tank car (pursuant to the January 25, 2011 *Federal Register* Notice) and the tank car proposed in AAR's petition.

Table 4. Comparison of Rail Tank Car Safety Features

Rail Tank Car Safety Features Comparison								
Rail Tank Car Specification	Bottom Outlet Handle	Gross Rail Load (lbs)	Head Shield Type[1]	Pressure Relief Valve	Shell Thickness	Tank Material	Top Fittings Protection	Thermal Protection System
DOT Specification 111	Attached During Transportation	263,000	Optional	STDP set 165 psig Optional Reclosing Requirement	7/16 inch Minimum	TC-128B or ASTM A516-70 Steel	Optional	Not Required; Optional
DOT Approved 286K Car (76 FR 4250)	Attached During Transportation	286,000	Half Height	STDP set 165 psig Optional Reclosing Requirement	7/16 inch-TC-128B or 1/2 inch-ASTM 516-70	Normalized TC-128B or ASTM 516-70 Steel	Yes	Not Required; Optional
AAR Petition 1577 (Non-Jacketed)	Attached During Transportation	286,000	Half Height	STDP Set 165 psig Reclosing Requirement	1/2 inch-TC-128B or 9/16 inch-ASTM 516-70	Normalized TC-128B or ASTM 516-70 Steel	Yes	Not Required; Optional
AAR Petition 1577 (Jacketed)	Attached During Transportation	286,000	Full Height[2]	STDP Set 165 psig Reclosing Requirement	7/16 inch-TC-128B or 1/2 inch-ASTM 516-70	Normalized TC-128B or ASTM 516-70 Steel	Yes	Not Required; Optional

[1] (1/2 inch Minimum Thickness)
[2] Full Height Head Shields Require a Jacket
STDP=Start to Discharge Pressure

Concurrent with completing the first draft of its ANPRM in May 2012, between April 2012 and October 2012, PHMSA received three additional petitions (P-1587, P-1595, and P-1612) and one modification of a previously filed petition (P-1612). These petitions were submitted by concerned communities and various industry associations requesting further modification to the tank car standards. In response to this additional information, PHMSA published an ANPRM on September 6, 2013,[11] which addressed all of the petitions and NTSB recommendations related to rail safety, including tank car and operational standards for flammable liquids.

Public interest in this rulemaking was significant. We received comments from local communities, cities and towns, rail carriers, shippers, equipment suppliers, tank car manufacturers, environmental groups, and the NTSB. PHMSA is reviewing the extensive public comments received during the comment period, which ended on December 5, 2013, and will use the comments to assess possible future regulatory changes. PHMSA, in coordination with FRA, is considering all regulatory avenues available to improve rail safety.

Tank cars are only one part of the chain of delivery, and we must identify and evaluate all of the risks associated with bulk movements of highly hazardous material, such as crude oil and ethanol, and then work to reduce or eliminate those risks.

In addition to the rulemaking activity by PHMSA, DOT took additional regulatory action following the Lac-Mégantic derailment. On August 7, 2013, FRA, in coordination with PHMSA, issued an emergency order[12] addressing the immediate hazard of death, personal injury, or significant harm to the environment, by instituting requirements related to attending and securing certain hazardous materials trains and cars, including crude oil and ethanol unit trains. The emergency order addressed the leading factors identified in preliminary findings in the Lac- Mégantic investigation, and PHMSA and FRA are conducting field inspections and investigations to monitor compliance with the emergency order.

On February 25, 2014, the Department issued another emergency order[13] to improve the safe transportation of crude oil by rail. The emergency order requires those who offer crude oil for transportation by rail to ensure the product is properly tested and classified in accordance with federal Hazardous Materials Regulations. All Class III crude oil shipments must now be designated as Packing Group I or II, thereby requiring the use of a more robust tank car. (Previously some Class III crude oil shipments were allowed to be designated as Packing Group III.) As our efforts in testing crude oil samples have uncovered evidence of misclassification, the emergency order highlights the importance of properly testing and classifying crude oil prior to shipping to ensure the product is being transported in containers that are designed to safely store the hazardous material while in transit.

Non-Regulatory Efforts

Concurrent with FRA's August 7, 2013 emergency order, PHMSA and FRA published a joint Safety Advisory[14] that addressed preliminary findings of the Lac-Mégantic investigation and made the following safety and security recommendations: (1) reminding railroads to review the adequacy of their crew staffing requirements for trains transporting hazardous materials; (2) requiring system-wide evaluations to identify particular hazards that may make it more difficult to secure a train or pose other safety risks; and (3) requiring that procedures be developed to mitigate those risks.

The joint Safety Advisory also announced an emergency meeting of FRA's Railroad Safety Advisory Committee (RSAC) to address rail safety concerns, which was held on August 29, 2013. [15] During the emergency meeting, PHMSA and FRA explained the safety requirements in the August 7, 2013 emergency order and the recommendations in the joint Safety Advisory, and proposed that an RSAC working group be formed to address hazardous materials transportation requirements. RSAC members discussed the formulation of a task statement regarding appropriate train crew size, hazard classes, and quantities of hazardous materials that should trigger additional operating procedures, including attendance and securement requirements. PHMSA continues to participate in FRA's RSAC meetings on hazardous materials transport by rail. The RSAC plans to provide its recommendations regarding hazardous materials rail safety by April 2014 to FRA, who will forward the recommendations to PHMSA for further evaluation.

In addition to participating in the RSAC meetings, PHMSA, as mentioned above, has been a participant in and an observer of the TCC. This committee is comprised of representatives of the Class I, short line, and regional railroads; rail tank car owners, manufacturers, and repair facilities; and shippers and customers of hazardous materials by rail, as well as participants from PHMSA, FRA, Transport Canada, and the NTSB. The TCC works together to develop technical standards for how tank cars, including those used to move hazardous materials, are designed and constructed. PHMSA also participates as a working member of other rail task forces.

On August 27-28, 2013, before the RSAC meeting, PHMSA and FRA held a public meeting to review the requirements in the Hazardous Materials Regulations applicable to rail operations.[16] PHMSA and FRA conducted this meeting as part of a comprehensive review of operational factors that affect the safe transportation of hazardous materials by rail. This meeting provided the opportunity for public input on requirements related to rail operations.[17] PHMSA and FRA are currently reviewing the transcript and public comments and will use the comments to inform their future possible regulatory changes.

On November 20, 2013, PHMSA and FRA issued another joint Safety Advisory to reinforce the importance of proper characterization, classification, and selection of a packing group for Class 3 materials (flammable liquids) and the corresponding regulations for safety and security planning.[18] This Safety Advisory noted that we expect offerors of hazardous material by rail and rail carriers transporting hazardous material to revise their safety and security plans as required under the Hazardous Materials Regulations, including the required risk assessments, to address the safety and security issues identified in FRA's August 7, 2013 emergency order and the August 7, 2013 joint Safety Advisory. FRA has initiated a focused effort to audit security plans, specifically at railroads that move unit trains of flammable liquids.

On January 2, 2014, PHMSA issued a Safety Alert warning of the variability in certain crude oil and emphasizing that proper and sufficient testing to ensure accurate characterization and classification should be performed.[19] Proper characterization and classification are integral for the Hazardous Materials Regulations to function as they were designed. Characterization and classification ultimately determine the appropriate and permitted packaging for a given hazardous material. This Safety Alert addressed the initial findings of “Operation Classification,” a compliance initiative (described below) involving unannounced inspections and testing of crude oil samples to verify that offerors of the materials have properly characterized and classified the hazardous materials. The Safety Alert expressed PHMSA’s concern that unprocessed crude oil may affect the integrity of the packaging or present additional hazards, related to corrosivity, sulfur content, and dissolved gas content. The alert also noted (1) that preliminary testing had focused on the classification and packing group assignments that have been selected and certified by offerors of crude oil, and (2) that PHMSA has found it necessary to expand the scope of its testing to measure other factors that might affect the proper characterization and classification of the materials.

Call to Action

On January 9, 2014, the Secretary issued a "Call to Action" to actively engage stakeholders in the crude oil and rail industries to take immediate steps to improve the transportation of crude by rail. On January 16, 2014, the Secretary held a meeting where the Administrators of PHMSA, FRA, and the Federal Motor Carrier Safety Administration challenged representatives of all stakeholders to identify prevention and mitigation strategies that can be implemented quickly.

Specifically, the "Call to Action" discussed (1) operational controls and track maintenance measures that could prevent accidents and (2) the proper classification and characterization of hazardous materials. The meeting was an open and constructive dialogue on how, collaboratively, industry and government can make America's railways and other modes of transportation for hazardous materials safer, since the misclassification of a hazardous material affects more than just the railroad industry. During the meeting, the rail and crude oil industries agreed to consider potential actions they could take to enhance safety, including speed restrictions in high-consequence areas, alternative routing, the use of distributive power to improve braking, increased track inspections, improvements to crude oil testing and classification processes, and emergency response preparedness and training. In addition, the participants agreed to return to the TCC for discussions on further improvements to the tank car standard. On January 22, 2014, the Secretary sent a letter to the attendees recapping the meeting and stressing the importance of this issue.[20]

In the weeks following the "Call to Action" meeting, PHMSA has worked closely with industry stakeholders to advance the safety initiatives. We have received voluntary agreements from the AAR, the American Short Line and Regional Railroad Association, and the American Petroleum Institute to consider ways to quickly implement the approaches to safety discussed during the meeting.

Enhanced Enforcement and Outreach

In addition to regulatory and non-regulatory efforts to improve rail safety, PHMSA has increased its efforts to improve awareness and understanding of, and compliance with, the Hazardous Materials Regulations. These efforts include enforcement and outreach activities that are focused on proper classification and characterization of hazardous materials, development of safety and security plans, and the awareness and understanding of Hazardous Materials Regulations. PHMSA has focused on addressing the considerable public, media, and congressional interest in the subject of crude oil transport by rail.

As mentioned above, PHMSA launched “Operation Classification,” a compliance initiative involving unannounced inspections and testing of crude oil samples to verify that offerors of the materials have properly classified and described the materials being shipped. In January 2013, PHMSA and FRA began planning this initiative and officially launched “Operation Classification” in August 2013. This initiative is an ongoing effort, and PHMSA will continue to collect samples and test them to determine the characteristics of Bakken crude oil, as well as oil from west Texas. To date, PHMSA has taken 58 samples to collect some preliminary information about the hazards associated with these oils. PHMSA has uncovered 11 potential violations, primarily related to improper packaging group assignment, and plans to release the findings of Operation Classification publicly upon conclusion of the effort. PHMSA has initiated enforcement actions on potential violators. In addition, as these violations could indicate further non-compliance issues, PHMSA continues to expand the scope of its investigations. PHMSA will use the results of the findings to consider the benefits, costs, and alternatives of any future regulatory action.

On January 17, 2014, PHMSA published a Web Page entitled "Operation Safe Delivery: Enhancing the Safe Transport of Flammable Liquids."[21] This site describes the Department's efforts to enhance the safe transport of flammable materials by rail and serves as a valuable resource for enhancing the safe transport of flammable liquids. The site will receive regular updates to provide progress reports on industry commitments as part of the “Call to Action” and additional Departmental activities related to rail safety initiatives. This site also displays the Department’s rail safety action plan. Although the site was only recently unveiled, it has already received considerable traffic and is an educational resource for industry and the general public.

PHMSA is also diligently responding to both congressional and media inquiries on the subject of crude oil transport by rail. Since the beginning of 2013, PHMSA has received and responded to over twenty letters from Members of Congress requesting information on this topic. Finally, PHMSA is coordinating with the Government Accountability Office on an audit of the transportation infrastructure of the United States used to accommodate increased shale oil and gas production.

III. Closing Remarks

During my four years as PHMSA's Administrator, I have experienced marked changes in our hazardous materials transportation landscape. The emergence of the United States as the world's leading energy producer has undoubtedly changed our transportation system and provided new challenges for PHMSA. I have seen these changes and the evolution of the energy industry firsthand. I have also seen the lasting consequences that transportation incidents can have on the public and local communities nearby. We must prepare for these new and shifting demands right now and ensure that we protect our communities and the environment. Effective standards and regulations are important mechanisms for keeping America's people and its environment safe while providing for the transportation of the Nation's energy supplies, and we will continue to use our authorities to improve the effectiveness of our standards and regulations. PHMSA's oversight and enforcement capabilities, along with those of our Federal and State partners, are critically important.

PHMSA and FRA are committed to improving the safety of the transportation of hazardous materials by rail and other modes. With this in mind, I believe that our comprehensive approach to rail safety is working, but we must continue to adapt our approach as we identify changing risks. Improvement in tank car integrity is one part of the ongoing effort to address the changes in the risks associated with transportation of hazardous materials. Furthermore, PHMSA and FRA are not alone in our safety efforts. As the "Call to Action" demonstrated, the rail and crude oil industries are integral partners in improving transportation safety, and PHMSA will continue to work collaboratively to improve safety. We have a long way to go to reach no deaths, injuries, environmental or property damage, or transportation disruptions, but I truly believe our efforts are helping prevent accidents and will help mitigate their damage.

In closing, we look forward to continuing to work with Congress to address rail safety issues, specifically those dealing with the bulk shipment of flammable liquids. Together, we will strive to keep America's people and its environment safe while providing for the reliable transportation of the Nation's energy supplies. Everyone at PHMSA is dedicated and committed to fulfilling our safety responsibility to the American people. It is an honor to serve the American people and to work with the dedicated public servants at PHMSA. Thank you again for the opportunity to speak with you today. I would be pleased to answer any questions you may have.

End Notes

[1] Data from the North Dakota Department of Mineral Resources Web site: https://www.dmr.nd.gov/oilgas/stats/historicalbakkenoilstats.pdf

[2] http://in.reuters.com/article/2014/01/16/usa-rail-regulator-idINL2N0KQ1WN20140116

[3] The Surface Transportation Board Web site indicates that the inflation-adjusted dollar amount for 2012 (the most recent year for which a complete year of annual operating revenue data is available) is $452,653,248 or more for a Class I railroad. Figures in table are quarterly totals. Source: Association of American Railroads

[4] Data from 2003-2012 compiled by FRA Office of Safety Analysis.

[5] See *Federal Register* http://www.gpo.gov/fdsys/pkg/FR-2011-01-25/pdf/2011-1414.pdf

[6] See *Federal Register* http://www.gpo.gov/fdsys/pkg/FR-2011-01-25/pdf/2011-1342.pdf

[7] See http://www.regulations.gov/#!documentDetail;D=PHMSA-2011-0059-0001

[8] "Packing Group" designates the hazard level posed by a class of materials. Class 3 (flammable liquids) Packing Group I materials have a low boiling point and represent a high flammability risk. Packing Group II materials have a higher boiling point and a low flash point and represent a slightly lower flammability risk.

[9] Table 4 provides a comparison of the DOT Specification 111 tank car currently authorized in the Hazardous Materials Regulations, the minimum standards for the DOT approved tank car pursuant to the January 25, 2011 *Federal Register* Notice and the tank car proposed for incorporation in petition (P-1577) by AAR.

[10] See NTSB recommendations: R-07-4, R-12-5, R-12-6, and R-12-7 http://www.phmsa.dot.gov/hazmat/regs/ntsb/rail

[11] See *Federal Register* http://www.gpo.gov/fdsys/pkg/FR-2013-09-06/pdf/2013-21621.pdf

[12] See *Federal Register* https://federalregister.gov/a/2013-19215

[13] See DOT website: http://www.dot.gov/briefing-room/dot-issues-emergency-order-requiring-stricter-standards- transport-crude-oil-rail

[14] See *Federal Register* https://federalregister.gov/a/2013-19211

[15] See *Federal Register* https://federalregister.gov/a/2013-19471

[16] See *Federal Register* https://federalregister.gov/a/2013-17201

[17] See public comments http://www.regulations.gov/#!docketDetail;D=FRA-2013-0067

[18] See *Federal Register* https://federalregister.gov/a/2013-27785

[19] See safety alert http://www.phmsa.dot.gov/staticfiles/PHMSA/DownloadableFiles/1_2_14%20Rail_Safety_Alert.pdf

[20] See Call to Action follow-up letter at http://phmsa.dot.gov/staticfiles/PHMSA/DownloadableFiles/Files/Letter_from_Secretary_Foxx_Follow_up_to_Ja nuary_16.pdf

[21] Web site available at http://phmsa.dot.gov/portal/site/PHMSA/menuitem.6f23687cf7b00b0f22e4c6962d9c8789/?vgnextoid=c5ff6d96d8283410VgnVCM100000d2c97898RCRD&vgnextchannel=0f0b143389d8c010VgnVCM1000008049a8c0RCRD&vgnextfmt=print

In: Transporting Crude Oil by U.S. Rail
Editor: Rosario S. McLaughlin
ISBN: 978-1-63117-837-5

Chapter 4

STATEMENT OF CHRISTOPHER A. HART, VICE CHAIRMAN, NATIONAL TRANSPORTATION SAFETY BOARD. HEARING ON "ENHANCING OUR RAIL SAFETY: CURRENT CHALLENGES FOR PASSENGER AND FREIGHT RAIL"*

Good afternoon, Chairman Blumenthal, Ranking Member Blunt, and Members of the Subcommittee. Thank you for the opportunity to appear before you on behalf of the National Transportation Safety Board (NTSB) and to update you on our ongoing work to improve railroad safety by investigating railroad accidents and issuing safety recommendations. Our nation's economy depends on a safe, reliable rail transportation system, and the American public expects and deserves nothing less. Recent railroad accidents under active investigation, including fatal accidents, remind us of the clear imperative to stay vigilant and stand ready to make improvements to the safety of railroad transportation. Our Nation's railroad system is safe, but evolving demands on the railroad system mean evolving safety challenges, and much work is ahead in our shared mission of making our nation's railroad system as safe as it can be.

---

* This is an edited, reformatted and augmented version of a statement presented March 6, 2014 before the Senate Committee on Commerce, Science, and Transportation, Subcommittee on Surface Transportation and Merchant Marine Infrastructure, Safety, and Security.

Recent events have placed railroad safety at the forefront of the national conversation. Last May, in Bridgeport, Connecticut, 76 people were injured when a Metro-North Railroad (Metro-North) commuter train derailed, fouled the adjacent track, and was struck by a train approaching on that adjacent track. Just more than a week later, a Metro-North track foreman was struck by a train and killed in West Haven, Connecticut. In July, a CSX train operating on Metro-North tracks derailed in The Bronx. In December, four people lost their lives and 59 others were injured when a Metro-North commuter train derailed in The Bronx after entering a curve with a 30-mile-per-hour (mph) speed limit at 82 mph. One month later, 2014 dawned with a team of NTSB investigators working the scene of a serious railroad accident near Casselton, North Dakota, where 20 cars of a 106-car BNSF petroleum crude oil unit train ignited after colliding with cars from a derailed BNSF grain train.[1] More than 476,000 gallons of crude oil were released in the accident, and the massive fire triggered a voluntary evacuation of 1,400 people from the surrounding area and resulted in millions of dollars in damage.

Our investigations into these accidents continue, and the second portion of this written testimony will update the Subcommittee on what we have learned so far. Last week, we issued three recommendations to Metro-North that Metro-North install signs to clearly warn train crews that they are approaching areas of permanent speed restrictions and that Metro-North install and review inward- and outward-facing audio and video recorders in locomotives and control cars, which is a longstanding NTSB recommendation to the Federal Railroad Administration (FRA).

Current Safety Issues

First, I would like to offer some perspectives on safety issues including, (1) safety deficiencies in the design of thousands of railroad tank cars; (2) the need for widespread implementation of positive train control (PTC) systems; (3) the need for installation of inward- and outward-facing locomotive cameras; and (4) the need for focused, industrywide efforts to foster top-down safety cultures in which safety thrives. There is not one approach that will improve rail safety, but it must be addressed by implementing varied approaches that, when working together, can help drive down the number and frequency of accidents or mitigate the severity of accidents.

Railroad Tank Car Design

The Nation's railroad network is taking on an expanding role—one that has profound economic importance—as a major channel for the transportation of crude oil and other hazardous products. As the NTSB noted recently, the American Association of Railroads' 2012 *Annual Report of Hazardous Materials Transported by Rail* states that crude oil traffic has increased by 443 percent since 2005 and that this growth is expected to continue for the foreseeable future. According to the FRA, the volume of crude oil transported by rail has increased dramatically in recent years, from approximately 65,600 carloads in 2011 to approximately 257,450[2] carloads in 2012—an increase of 292 percent.[3] Moreover, not only is more crude oil being transported by rail, but some of the crude oil being moved on the Nation's railroad system—such as that originating in the Bakken formation—may have more volatile properties. Last month, the Pipeline and Hazardous Materials Safety Administration (PHMSA) issued a safety alert advising "the general public, emergency responders and shippers and carriers that ... the type of crude oil being transported from the Bakken region may be more flammable than traditional heavy crude oil," with the results of further tests of Bakken crude oil forthcoming.[4]

Furthermore, ethanol traffic transported by railroad increased 442 percent between 2005 and 2010; in 2012, ethanol was the most frequently transported hazardous material in the railroad system.[5] The evolving role of our Nation's railroad network in the transportation of flammable crude oil and ethanol requires interested parties to take a comprehensive approach to eliminate or significantly reduce the safety risks. This approach must include improvements to track inspection and maintenance programs and the crashworthiness of the tank cars that transport these materials.

Indeed, as the volume of flammable materials transported by rail grows the Casselton, North Dakota, accident has become an increasingly commonplace story—and multiple recent serious and fatal accidents reflect substantial shortcomings in tank car design that create an unacceptable public risk. The crude oil unit train involved in the Casselton accident consisted of railroad tank cars designed and manufactured to Department of Transportation (DOT) Specification 111-A100W1 (DOT-111)—a design that presents demonstrated and serious safety concerns when used to transport hazardous materials such as crude oil. Specifically, the NTSB has identified vulnerabilities in DOT-111 tank car design with respect to tank heads, shells, and fittings that create the unnecessary and demonstrated risk that, in an accident, hazardous materials could be released and, in the case of flammable materials such as crude oil and ethanol, could ignite and cause catastrophic damage.[6]

The NTSB continues to find that accidents involving the rupture of DOT-111 tank cars carrying hazardous materials often have violent and destructive results. For example, on July 6, 2013, a 4,700-foot-long train that included 72 DOT-111 tank cars loaded with crude oil from the Bakken fields derailed in Lac-Mégantic, Quebec, triggering an intense fire fed by crude oil released from at least 60 cars. The fire engulfed the surrounding area and completely destroyed buildings and property. Forty-seven people died. The NTSB is assisting the Transportation Safety Board of Canada (TSB) in its investigation of that accident, and last month both the NTSB and TSB issued safety recommendations asking FRA and PHMSA, as appropriate, to require railroads to evaluate the safety and security risks of crude oil train routes and select routes that avoid populous and other sensitive areas, require railroads to develop comprehensive emergency response plans for worst- case releases resulting from accidents, and require shippers to sufficiently test and properly classify hazardous materials such as crude oil prior to shipment. With respect in particular to the recommendation on classification of crude oil, PHMSA's safety alert regarding Bakken crude oil underscores the critical importance of accurate classification of flammable materials. We look forward to working with PHMSA and FRA on implementing these recommendations.

In addition, the NTSB is investigating, or has investigated, a spate of recent similar accidents in the United States that demonstrate the destructive results when DOT-111 tank cars containing hazardous materials are punctured, including:

- The July 11, 2012, Norfolk Southern Railway Company train derailment in a Columbus, Ohio, industrial area in which three derailed DOT-111 tank cars released about 54,000 gallons of ethanol, with energetic rupture of one tank car in a post-accident fire.
- The October 7, 2011, Tiskilwa, Illinois, train derailment of 10 DOT-111 tank cars resulting in fire, energetic rupture of several tank cars, and the release of 162,000 gallons of ethanol.
- The June 19, 2009, Canadian National Railway train derailment in Cherry Valley, Illinois, in which 13 of 19 derailed DOT-111 tank cars were breached, caught fire, and released about 324,000 gallons of ethanol. The post-accident fire resulted in one death, nine injuries, and the evacuation of 600 houses within half a mile of the accident.

- The October 20, 2006, New Brighton, Pennsylvania, Norfolk Southern Railway Company train derailment in which 23 DOT-111 tank cars derailed, fell from a bridge, caught fire, and released more than 485,000 gallons of ethanol.

Federal requirements simply have not kept pace with evolving demands placed on the railroad industry and evolving technology and knowledge about hazardous materials and accidents. In fact, the current American Association of Railroads (AAR) industry standards adopted for DOT-111 tank cars ordered after October 1, 2011 that are used to transport packing group I and II crude oil, impose a level of protection greater than corresponding federal requirements. [7] At this point, however, the NTSB is not convinced that these modifications offer significant safety improvements.

The NTSB continues to assert that DOT-111 tank cars, or tank cars of any successor specification, that transport hazardous materials should incorporate more effective puncture-resistant and thermal protection systems. This can be accomplished through the incorporation of additional protective features such as full head shields, jackets, thermal insulation, and thicker head and shell materials. Because the average service life of a tank car may run 20-30 years, it is imperative that industry, the FRA, and PHMSA take action now to address hazards that otherwise would exist for another half- generation or longer.

Following the 2011 Cherry Valley, Illinois, accident the NTSB recommended that PHMSA improve DOT-111 tank car crashworthiness by:

> [r]equir[ing] that all newly manufactured and existing general service tank cars authorized for transportation of ... crude oil in Packing Groups I and II have enhanced tank head and shell puncture-resistance systems and top fittings protection that exceeds existing design requirements for DOT-111 tank cars.[8]

The NTSB also recommended that PHMSA improve requirements for bottom outlet valves so that they remain closed during accidents involving impact forces[9] and require improved center sill or draft sill attachment designs,[10] and the NTSB reiterated its prior recommendation that PHMSA, in consultation with FRA, require that railroads immediately provide emergency responders with accurate, real-time information on hazardous materials on a train.[11]

The importance of providing correct information to first responders highlights a related issue. Following the freight train derailment in Paulsboro, New Jersey, on November 30, 2012, which is the subject of an ongoing NTSB investigation, the NTSB learned of the critical importance to first responders of immediate, accurate information about the contents of a derailed tank car so that first responders may tailor their emergency response in a manner that best protects life and property. First responders' ability to make good decisions in responding to a hazardous-materials release depends on their clear understanding of what is in a tank car. Any improvement to railroad tank car safety must proceed hand-in-hand with an improved approach to ensuring first responders have adequate information to take appropriate life-saving actions. PHMSA indicates it, along with FRA, is working to implement this recommendation.

Although important decisions are clearly ahead for regulators and industry, the NTSB is pleased that at least some progress has been made. PHMSA published an advance notice of proposed rulemaking (ANPRM) on September 6, 2013, for potential safety improvements to DOT-111 tank cars, and we remain engaged in that rulemaking proceeding. In NTSB comments on the ANPRM dated December 5, 2013, we urged PHMSA to promptly address the four recommendations that were included in the NTSB report on the Cherry Valley accident, described above, and to issue improved and effective regulations that reduce the risks associated with DOT-111 tank cars. We will continue to carefully monitor PHMSA's progress and will ensure decision-makers have the full benefit the lessons the NTSB has learned through its investigations. The NTSB also continues to call on industry stakeholders to rise to the challenge and explore measures that will improve tank car design in the interim. Industry and Department of Transportation leaders met in January to discuss development of an industry plan to make certain safety improvements.[12] This dialogue is encouraging, and the NTSB will continue to urge regulators and industry stakeholders to follow through on any commitments.

Implementation of PTC Systems

PTC systems help prevent (a) derailments caused by overspeeding, (b) train-to- train collisions by slowing or stopping trains that are not being operated in accordance with the signal systems and operating rules, and (c) injury to track workers. The first NTSB-investigated accident that train control technology would have prevented occurred in 1969, when four people died and 43 were injured in the collision of two Penn Central commuter trains in Darien, Connecticut. The NTSB recommended in response to that accident that FRA study the feasibility of requiring railroads to install an automatic train control system, the precursor to today's PTC systems.[13]

More recently, in 2008, more lives were lost in a PTC-preventable accident when a Metrolink commuter train and a Union Pacific freight train collided head-on in Chatsworth, California, killing 25 people and injuring 102 others. The NTSB concluded that the Metrolink engineer's use of a cell phone to send text messages distracted him from his duties. PTC would have prevented the tragedy that resulted. In the aftermath of the Chatsworth accident, Congress enacted the Rail Safety Improvement Act (RSIA) of 2008, which requires each class I rail carrier and each provider of regularly scheduled intercity passenger or commuter rail transportation to implement a PTC system by December 31, 2015, on each line over which intercity passenger or commuter service is operated or over which poison- or toxic-by-inhalation hazardous materials are transported.[14] In 2012, however, FRA exempted about 10,000 miles of track from the PTC mandate, and several rail carriers and transit authorities have stated that they will not meet the 2015 deadline.

We continue to see accidents that could be prevented by PTC. The December 1 Metro-North accident in the Bronx, which killed four people and injured 59 others, would have been prevented by PTC. We also are examining the role PTC could have played in the May 28 roadway worker fatality. Since 2004 alone, in the 25 PTC-preventable freight and passenger rail accidents that NTSB investigated, 65 people died, more than 1,100 were injured, and damages totaled millions of dollars.[15] With each PTC-preventable accident, the case for PTC only grows stronger, yet progress toward industry-wide implementation has been slow.

Implementation of PTC systems was included on the NTSB's Most Wanted List when the list was first published in 1990 and has remained on the list almost continuously since that time. We may never eliminate human error from the railroad system, but PTC provides a level of redundancy to protect trains and those on board when human factors, such as distraction or fatigue, might otherwise set an accident sequence into motion.

Some rail carriers have installed PTC or are working to meet the 2015 deadline. However, in August 2013, the Government Accountability Office reported to the U.S. Senate that, due to a number of complex and interrelated challenges, the majority of railroads will not complete PTC implementation by the 2015 deadline.[16] NTSB files are filled with accidents that could have been prevented by PTC, and for each and every day that PTC implementation is delayed, the risk of an accident remains. For PTC to reach its greatest safety potential, it must be implemented on all passenger and freight lines subject to the installation requirements specified in the RSIA. Lives depend on it.

There is much debate by policymakers over whether to extend the 2015 deadline established by RSIA. If Congress were to delay the statutory deadline, railroads that had delayed planning PTC implementation would be rewarded and railroads that had moved ahead with planning for PTC implementation by the deadline would essentially be punished. If the deadline remains unaltered, the NTSB would encourage FRA to take appropriate action to ensure railroads are complying with all applicable requirements regarding PTC implementation.

Following the head-on collision of two Union Pacific freight trains in Goodwell, Oklahoma, NTSB issued the following recommendation to the railroads covered under the RSIA mandate:

> Provide positive train control implementation update reports to the Federal Railroad Administration every 6 months until positive train control implementation is complete. The update reports should consist of two sections: components and training. The components section should include a description of the positive train control component to be implemented, the number of components, the number of components completed on the report date, the number of components that remain to be completed, the overall completion percentage, and the estimated completion date. Components are defined as locomotives, wayside units, switches, base station radios, wayside radios, locomotive radios, and any new and novel technologies that are part of a positive train control system. The training section should include the number of safety-related employees and equivalent railroad carrier contractors and subcontractors that need to be trained, by class and craft; minimum training standards for those employees and contractors, meaning the knowledge of and ability to comply with federal railroad safety laws and regulations and carrier rules and procedures to implement positive train control; the percentage of employees who have completed training; the percentage of employees who remain to be trained; and the estimated date that training will be completed.

Additionally, the NTSB recommended that FRA:

> Publish the positive train control implementation update reports submitted by all railroads subject to the positive train control provisions

> of the Rail Safety Improvement Act of 2008 and make the reports available on your website within 30 days of report receipt.[17]

The NTSB believes this information should be made available online to ensure a transparent accounting for actions taken and not taken to meet the 2015 deadline so that regulators and policymakers can make informed decisions. However, because of FRA's lack of sufficient action on its recommendation, this week we classified this recommendation as "Open—Unacceptable Response."

Inward- and Outward-Facing Locomotive Audio and Image Recorders

The December 1, Metro-North accident in The Bronx raised questions about the actions of the engineer prior to the crash. The NTSB has repeatedly called for railroad carriers to install inward- and outward-facing audio and image records to answer similar questions that have arisen in other accidents. Recorders in locomotives and cab car operating compartments are critically important not only because they would assist NTSB investigators and others understand what was happening in a train in the minutes and seconds before an accident, but also because they would help railroad management *prevent* accidents by identifying and responsibly addressing safety issues before they lead to injuries and loss of life and allow for the development of material that can be a valuable training and coaching tool.

The Chatsworth tragedy in 2008 again made the case crystal-clear for understanding crewmembers' activities in the minutes and seconds leading up to accidents. Discussing the strong safety case for a requirement for inward-facing cameras in locomotives, the NTSB noted that

> [i]n all too many accidents, the individuals directly involved are either limited in their recollection of events or, as in the case of the Chatsworth accident, are not available to be interviewed because of fatal injuries. In a number of accidents the NTSB has investigated, a better knowledge of crewmembers' actions before an accident would have helped reveal the key causal factors and would perhaps have facilitated the development of more effective safety recommendations.

On the basis of that reasoning, the NTSB recommended that FRA require the installation, in control compartments, of "crash- and fire-protected inward-

and outward-facing audio and image recorders capable of providing recordings [for at least 12 hours] to verify that train crew actions are in accordance with rules and procedures that are essential to safety as well as train operating conditions."[18] The NTSB also recommended that FRA "[r]equire that railroads regularly review and use in-cab audio and image recordings ... to verify that train crew actions are in accordance with rules and procedures that are essential to safety."[19]

The NTSB recently reiterated these important recommendations in its report on the collision of a BNSF coal train with the rear end of a standing BNSF maintenance-of- way equipment train near Red Oak, Iowa, which resulted in fatal injuries to the two crewmembers of the striking train. Damage was in excess of $8.7 million. As the NTSB stated in its report, the accident again demonstrated the need for in-cab recording devices to better understand (and thereby prevent) serious railroad crashes that claim the lives of crewmembers, passengers, and the public.

Last week, we issued our longstanding recommendation on this subject directly to Metro-North Railroad. An industry-wide FRA-mandated approach would be far more effective, but failing that, we will address the recommendation on an individual basis.

Safety Culture

Fostering the development of transparent, top-to-bottom safety cultures in transportation is an important priority of the NTSB. Creating and nurturing a thriving safety culture within rail carriers is even more imperative in light of the expanding role of the Nation's railroad system as a main transporter of flammable materials.

The NTSB held a public forum on September 10 and 11, 2013, on successes and challenges associated with creating and maintaining strong safety cultures across the transportation modes, including rail. Panels of experts from academia, industry, and Federal regulatory agencies, such as the FRA, offered their perspectives on the significant organizational commitments and managerial work that are required to maintain safety cultures across large, complex organizations such as transportation carriers.

As Members of the Subcommittee well remember, organizational factors at the Washington Metropolitan Area Transit Authority (WMATA) contributed to the fatal June 22, 2009, Metrorail train collision near the Fort Totten station in Washington, D.C. The NTSB found that WMATA leaders did not take sufficient action to make safety a priority and to identify and address safety issues from the top down: the WMATA General Manager did not provide adequate information about critical safety issues; the WMATA Board of Directors did not seek information about critical safety issues; and the Board of Directors did not exercise oversight responsibility for system safety.

While WMATA has addressed many of these issues, the NTSB is examining the role of safety culture in the Metro-North accident investigations. The NTSB public investigative hearing regarding last May's two Metro-North accidents examined the importance of an organizational safety culture and the critical role that organizational culture plays in preventing accidents. The NTSB learned during the hearing that Metro- North has undertaken efforts to foster a stronger safety culture but that challenges remain. The then-president of Metro-North, spoke of the "challenge" associated with creating "a clear understanding across the organization that safety is the core value and we're not looking for shortcuts, and that we want people to work safely."[20] Metro-North officials and labor stakeholders further assured the NTSB that Metro-North is taking action to address safety issues from the top to the bottom. Organizational issues within the railroad industry will continue to be an area of examination as the NTSB investigations of the Bridgeport, West Haven, and two Bronx accidents continue. In particular, the Bridgeport accident underscores the critical importance of regular, vigorous, and robust inspections of tracks. Railroad management must afford track workers adequate time and opportunity to conduct inspections and make repairs as necessary. As part of its ongoing investigation, the NTSB is undertaking a comprehensive review of Metro-North track inspections and follow-up work and is also looking at the adequacy of the FRA's Track Safety Standards.

The importance of building relationships between management and employees that foster a vibrant safety culture cannot be overlooked. Trust is an essential ingredient in those relationships. A culture in which front-line employees may openly report operational errors and safety issues without fear of reprisal is absolutely critical, and, as we have seen in the aviation context, improves safety.

The NTSB will continue to urge Federal regulators, such as FRA and the Federal Transit Administration (FTA), to facilitate establishment of appropriate safety cultures. The WMATA accident, in particular, underscored the critical need for rail mass transit operators to enhance and nourish safety cultures. Our 2014 Most Wanted List reaffirms our view that:

> [t]he FTA should consider the elements of safety culture, crew resource management, fatigue risk management, and technology, as well as lessons learned from the rail industry, as it moves forward with [new legislative authority to set and enforce new safety standards and conduct investigations]. Identifying and implementing these will be key to saving lives and preventing injuries.

UPDATES ON ONGOING INVESTIGATIONS

I would now like to update the Subcommittee on developments in several recent investigations that are keeping the NTSB's railroad investigators very busy and demonstrate the need for continued vigilance in the railroad operating environment.

Metro-North Railroad Accidents

On December 1, 2013, the NTSB launched an investigative team to The Bronx, where a Metro-North Railroad commuter train with approximately 115 passengers on board derailed shortly after 7:15 a.m. near the Spuyten Duyvil station, while going from Poughkeepsie to Grand Central Station in New York City. Four passengers died in this Thanksgiving holiday weekend accident and at least 59 others sustained injuries requiring medical treatment. Metro-North estimated damage in excess of $9 million.

The NTSB issued a preliminary factual report on the accident on January 14, 2014. Based on evidence obtained to this point, our investigators have found that the accident train was traveling at approximately 82 mph when it derailed in the curve on approach to the Spuyten Duyvil station. The speed limit for the curve was 30 mph. PTC would have prevented this accident. Metro-North does not have a PTC system and has stated it will not meet the deadline, but it thought its system would be as robust as PTC. This accident clearly demonstrated that it was not.

Investigators have conducted detailed inspections and testing of the signal system, train brakes, and other mechanical equipment, and thus far have found no anomalies. They found no pre-accident anomalies in the track in the derailment area. Investigators have interviewed the train crewmembers, including the engineer and first responders and will continue to obtain and examine evidence from NTSB headquarters as the investigation proceeds. We are receiving excellent cooperation from the parties to the investigation: the FRA, Metro-North, the New York Public Transportation Safety Board, Teamsters Local 808, the New York Police Department, the New York Fire Department, and Bombardier Transportation, which manufactured the cars involved in the accident.

Last week, as I stated earlier, the NTSB issued three safety recommendations to Metro-North. We recommended that Metro-North install signs to clearly warn train crews that they are approaching areas of permanent speed restrictions; install inward- and outward-facing audio and video recorders in locomotives and control cars, and regularly review and use in-cab audio and image recordings in conjunction with other performance data, to verify train crew actions are in accordance with rules and procedures that are essential to safety. As our investigation proceeds, the NTSB will be prepared to issue additional safety recommendations if we determine any further safety improvements are necessary pending completion of our investigation.

The Bronx derailment was the fourth accident involving Metro-North property to trigger an NTSB investigation in 2013. Last June, the NTSB issued an urgent safety recommendation following the May 28 accident in which the track foreman died, calling upon Metro-North to immediately implement redundant measures to ensure the safety of track workers[21] (Metro-North indicates it is implementing this recommendation), and the NTSB reiterated a safety recommendation to the FRA to require redundant signal protection, such as shunting, for maintenance-of-way work crews who depend on the train dispatcher to provide signal protection.[22] That recommendation is currently classified as open with an acceptable response.

I am pleased to inform the Subcommittee that Metro-North has fully cooperated in all these investigations, at a difficult time for the railroad and its employees in the wake of several closely-spaced accidents. We anticipate and look forward to Metro- North's continued cooperation as the investigations proceed.

Although it is still too early in our investigations of these accidents to draw definitive conclusions, we will seek answers to the following questions, among others: What caused these accidents? Are there common threads among the accidents? What improvements can Metro-North, regulators, and others adopt that will prevent similar accidents from occurring in the future?

The NTSB tentatively expects to complete our investigations of all four accidents involving Metro-North trains or property in the second half of this year. We recognize and share the Subcommittee's sense of urgency to understand what lessons can be drawn from these accidents and to ensure that the railroad industry and its regulators implement appropriate safety improvements to prevent recurrences, and if our investigation reveals problems that need immediate attention, we will not hesitate to issue appropriate recommendations before we complete the investigation.

Freight Train Collision and Crude Oil Release near Casselton, N.D.

As I noted above, the NTSB is investigating the December 30, 2013, Casselton, North Dakota, accident that resulted in a significant post-crash fire that triggered a voluntary evacuation of about 1,400 people from the surrounding community.

The accident sequence began shortly after 2 p.m. when 13 cars of a 112-car westbound BNSF Railway grain train derailed. One of the derailed cars came to rest on the adjacent track. Shortly afterward, a BNSF 106-car petroleum crude oil unit train travelling east on the parallel track collided with the derailed grain car. The collision caused the head-end locomotives and the first 21 cars of the crude oil train to derail. Some of the crude oil tank cars were punctured during the accident releasing crude oil that ignited and caused the energetic rupture of several other tank cars. Dense smoke and concern over expanding fires resulted in voluntary evacuation of the surrounding area.

The crews on the two trains were uninjured. No injuries to the public were reported. Damage was estimated at $6.1 million.

On a preliminary basis, we have found that, of the 20 tank cars that derailed, 18 tank cars were breached and more than 476,000 gallons of crude oil were released. NTSB investigators have completed the on-scene portion of the investigation, including interviews with the train crews and first responders. A broken axle and two wheels were shipped to the NTSB materials laboratory in Washington, D.C., for further evaluation and analysis, as well as the locomotive event and video recorders. The parties to the investigation include the FRA; PHMSA; the BNSF; the Brotherhood of Locomotive Engineers and Trainmen; the International Association of Sheet Metal, Air, Rail and Transportation Workers, formerly known as the United Transportation Union; Trinity Rail Car; and Standard Steel, LLC.

CONCLUSION

Thank you for the opportunity to appear before you and to provide updates on our ongoing investigations as well as NTSB perspectives on several compelling safety issues. Please be assured that the NTSB will remain engaged on these and all issues affecting transportation safety. I look forward to answering the Subcommittee's questions.

End Notes

[1] A unit train is a train made up of cars carrying the same product.

[2] The capacity of a tank car is about 30,000 gallons or 675 barrels of oil.

[3] FRA Emerg. Order No. 28, 78 Fed. Reg. 48218, 48220 (Aug. 7, 2013).

[4] PHMSA Safety Alert: Preliminary Guidance from Operation Classification (Jan. 2, 2014).

[5] FRA Emerg. Order No. 28, 78 Fed. Reg. at 48221; *see also*NTSB, Letter to The Honorable Cynthia L. Quarterman, Administrator, Pipeline and Hazardous Materials Safety Administration, U.S. Department of Transportation (Jan. 21, 2014), at 7 n. 11-13 (and citations therein).

[6] *See, e.g.,* NTSB, *Derailment of CN Freight Train U70691-18 With Subsequent Hazardous Materials Release and Fire Cherry Valley, Illinois, June 19, 2009,* Accident Rpt. No. NTSB/RAR-12/01 (Feb. 14, 2012), at 88 (concluding that, in accident involving breaches of DOT-111 tank cars, "If enhanced tank head and shell puncture-resistance systems such as head shields, tank jackets, and increased shell thicknesses had been features of the DOT-111 tank cars involved in this accident, the release of hazardous materials likely would have been significantly reduced, mitigating the severity of the accident.").

[7] These new standards, for example, call for DOT-111 tank cars that transport flammable liquids in packing groups I and II (the highest-risk of the three packing groups, classified according to flash and boiling points) to be built with protective "jackets" around their tanks, constructed of normalized steel at least 7/16 inch thick, and call for non-jacketed tanks to be constructed from normalized steel (steel that has been subjected to a heat-treating process that improves its material properties) at least half an inch thick. *See* American Assoc. of Railroads, *Manual of Standards and Recommended Practices: Specifications for Tank Cars,* M-1002. Corresponding Federal regulations require steel thickness of at least 7/16 inch, but they allow for the use of non-normalized steel and do not require incorporation of jackets or head shields. *See* 49 C.F.R. part 179, subpart D.

[8] *See* NTSB, *Derailment of CN Freight Train U70691 With Subsequent Hazardous Materials Release and Fire, Cherry Valley, Illinois, June 19, 2009*, RAR-12/01 (2012); NTSB Recommendation No. R-12-5 (2012).

[9] NTSB Recommendation No. R-12-6

[10] NTSB Recommendation No. R-12-7

[11] NTSB Recommendation No. R-07-4.

[12] "Rail, Oil Industries Weigh New Safety Measures in Wake of Derailments, Explosions," NBC News Investigations, Jan. 16, 2014, *available at* http://investigations.nbcnews.com/_news/2014/01/16/22328508- rail-oil-industries-weigh-new-safety-measures-in-wake-of-derailments-explosions?lite.

[13] NTSB Recommendation No. R-70-020.

[14] Rail Safety Improvement Act of 2008, Pub. L. No. 110-432, § 104 (2008).

[15] These accidents do not include Metro-North accidents.

[16] Gov't Accountability Office, *Positive Train Control: Additional Authorities Could Benefit Implementation*, GAO Rpt. No. GAO-13-720 (August 2013), *available at* http://www.gao.gov/assets/660/656975.pdf.

[17] *See* NTSB, *Head-On Collision of Two Union Pacific Railroad Freight Trains Near Goodwell, Oklahoma June 24, 2012,* Rpt. No. NTSB/RAR-13/02 (June 18, 2013); Recommendations Nos. R-13-23 and R-13-27 (2013).

[18] NTSB Recommendation No. R-10-1.

[19] NTSB Recommendation No. R-10-2.

[20] NTSB, Board of Inquiry in the Matter of Two Metro-North Rail Accidents: Bridgeport Train Derailment on May 17, 2013 and West Haven Collision Death of a Metro-North Track Foreman on May 28, 2013, Tr. Vol. II at 329 (Washington, D.C., Nov. 7, 2013).

[21] NTSB Recommendation No. R-13-17.

[22] NTSB Recommendation No. R-08-6.

In: Transporting Crude Oil by U.S. Rail
Editor: Rosario S. McLaughlin

ISBN: 978-1-63117-837-5

Chapter 5

STATEMENT OF GEOFFREY C. BLACKWELL, CHIEF, OFFICE OF NATIVE AFFAIRS AND POLICY, FEDERAL COMMUNICATIONS COMMISSION. HEARING ON "ENHANCING OUR RAIL SAFETY: CURRENT CHALLENGES FOR PASSENGER AND FREIGHT RAIL"*

Chairman Blumenthal, Ranking Member Blunt, and Members of the Subcommittee, thank you for this opportunity to appear before you today concerning the Federal Communications Commission's role in enhancing rail safety. My name is Geoffrey Blackwell, and I am the Chief of the Federal Communications Commission's (FCC) Office of Native Affairs and Policy, or ONAP, which is part of the Consumer and Governmental Affairs Bureau. ONAP was established in August 2010 to promote the deployment and adoption of communications services and technologies throughout Tribal Lands and Native Communities and to ensure robust government-to-government consultation with federally-recognized Tribal Governments and increased coordination with Native Organizations, including on environmental and historic preservation reviews and related matters. Along with other Commission Bureaus and Offices, ONAP is part of the internal infrastructure

* This is an edited, reformatted and augmented version of a statement presented March 6, 2014 before the Senate Committee on Commerce, Science, and Transportation, Subcommittee on Surface Transportation and Merchant Marine Infrastructure, Safety, and Security.

team at the Commission that coordinates closely on the matters involved in this hearing.

The Commission recognizes that positive train control (PTC) is a transformative technology that has the power to save lives, prevent injuries, and avoid extensive property damage. Clearly, PTC deployment is a top priority of all of the agencies and companies represented at the table today. It is also a top priority of the Commission to work with all parties to help them fulfill their various legal obligations and responsibilities and advance the deployment of PTC within the timeframe prescribed by the Railroad Safety Improvement Act of 2008 (RSIA).

At the same time, we recognize the need to balance legitimate concerns regarding the potential impacts of such a major infrastructure deployment on historic properties, including sites of religious and cultural significance to Tribal Nations. This imperative is reflected in the FCC's statutory obligations under Section 106 of the National Historic Preservation Act (NHPA), as well as the National Environmental Policy Act (NEPA). Most notably, NHPA regulations developed by the Advisory Council on Historic Preservation (ACHP) require that the FCC review and assess whether a proposed communications facility may have an adverse effect on historic properties, and consult with Federally recognized Tribal Nations as part of its review process.

For example, railroad tracks often follow historical trade and navigation routes that have been used for centuries, both by European settlers and by Native peoples before and since European settlement. Construction along the tracks therefore has the potential to affect a number of cultural or religious properties, from traditional burial sites to sacred landscapes. Our consultations with Tribal Nations revealed such specific histories. According to the Tribal Historic Preservation Officer (THPO) of the Quapaw Tribe of Oklahoma, when the railroad was constructed through historical Quapaw community lands in the southern plains, the soil of multiple burial mounds was bulldozed into the right-of-way to build up the track bed. According to the THPO of the Mississippi Band of Choctaw Indians, the largest Tribal cemetery ever discovered in the state of Mississippi was found under a railroad right-of-way. Several THPOs for Tribal Nations resident in both North and South Dakota have identified village sites that lie underneath existing rail lines. Likewise, Tribal Nations with an interest in the Columbia River Gorge have identified sacred sites and petroglyphs that are either underneath or alongside the existing tracks. Construction along railroad tracks can also affect historic properties that are not of cultural and religious significance to Tribal Nations. For example, the Pennsylvania, Georgia, Montana, Wyoming, and New

Mexico SHPOs have identified historic districts in their states that lie astride rail lines.

Accordingly, to address these concerns the Commission is dedicating substantial resources to establishing a specially-designed review process that would allow the railroads to deploy PTC wayside infrastructure by the December 31, 2015, statutory deadline, while also ensuring that it meets its statutory obligation to fully consider the effects of this infrastructure on historic properties, including sites of religious and cultural significance to Tribal Nations.

BACKGROUND

RSIA requires that freight and passenger railroads deploy by December 31, 2015, interoperable PTC systems along routes transporting passengers and certain hazardous cargoes. While the Federal Railroad Administration (FRA) is responsible for overseeing the railroads' implementation of PTC by the statutory deadline, the FCC plays a role in two aspects of PTC deployment. First, the FCC regulates access to spectrum, and many railroads are installing PTC systems that require licensed spectrum to enable wireless communication. Second, it is well established—and has been upheld by the D.C. Circuit—that a Commission licensee's construction of a facility (e.g., a communications tower) to use Commission-licensed spectrum constitutes a major federal action and federal undertaking within the meaning of NEPA and the NHPA. Thus, the FCC has a legal responsibility to review the environmental and historic preservation impacts, including impacts on properties of religious and cultural significance to Tribal Nations, of infrastructure projects, like PTC, that will provide wireless services using FCC-licensed spectrum. It is a common misperception that the responsibility to ensure that Tribal Nations have an opportunity to participate in this process arises only with respect to the lands upon which Tribal Nations currently reside. In fact, the NHPA requires a nationwide scope, and Tribal involvement in the FCC process is typically most important on non-Tribal lands. These previously inhabited areas are precisely where Tribal governments do not presently own the land or have the authority to protect their religious and cultural properties, except through the NHPA.

PROVIDING SPECTRUM

Broadly speaking, the FCC understands that two types of PTC systems will be implemented in the United States. While they both use frequencies in the 217 to 222 megahertz range, they have different spectrum requirements. In the Northeast Corridor, Amtrak and the commuter railroads intend to implement Amtrak's Advanced Civil Speed Enforcement System (ACSES), via 12.5 kHz spectrum channels. Outside of the Northeast Corridor, commuter railroads, Amtrak and the freight railroads intend to implement a system developed by the freight railroads, Interoperable Electronic Train Management System (I-ETMS), via 25 kHz channels.

Since 2010, FCC staff has been working with freight and commuter rails to identify spectrum that is suitable and available for the deployment of an interoperable PTC system. PTC-220, a consortium of the nation's seven Class I freight railroads, has acquired a substantial block of nationwide spectrum, as well as additional spectrum in dense urban markets, for the purpose of implementing PTC systems. FCC staff continues to work with PTC-220 and other relevant stakeholders to enable more intensive use of PTC-220's spectrum holdings, which will allow other railroads, especially commuter railroads, to benefit from PTC-220's spectrum holdings and networks. Provided that pending and future FRA requirements do not materially impact PTC spectrum needs, we believe the seven Class I freight railroads generally have sufficient spectrum to deploy PTC. We also believe that an increasing number of commuter rails have access to sufficient spectrum to deploy PTC. We continue to work actively with those still seeking spectrum.

ENVIRONMENTAL AND HISTORIC PRESERVATION REVIEW

The Commission's rules require that applicants ascertain—prior to initiating construction— whether a facility they wish to construct could have a significant environmental impact, including an adverse effect on historic properties. If the facility may have such an impact, the Commission's rules require the preparation of an Environmental Assessment (EA) and further environmental processing.

To determine whether there will be an adverse effect on historic properties, the rules require the applicant to follow certain procedures set forth in the Commission's 2004 Nationwide Programmatic Agreement (NPA) with ACHP and the National Conference of Historic Preservation Officers (NCSHPO). Intended to facilitate historic preservation review for communications projects, the NPA includes procedures for participation of Tribal Nations and, where necessary, referral to the Commission for government-to-government consultation. The NPA also improved the process for SHPO review and excluded some constructions from review under defined conditions. The Commission's rules also specify procedures for NEPA compliance. No formal regulatory action is required to adapt these procedures for PTC wayside poles, and the Commission staff is working with the railroads to make the necessary informal adjustments to the process.

While the Commission's rules include certain exemptions and exclusions from NEPA and NHPA review, none currently available would categorically exclude the entire PTC wayside pole system. The creation of exemptions for infrastructure deployments like PTC would require not only consultation and negotiation with ACHP and NCSHPO, the other necessary signatories to our NPA under ACHP rules, but also a full notice and comment rulemaking proceeding at the Commission, a process that could take a year or longer.

A cornerstone of the Commission's implementation of the NPA is the Tower Construction Notification System (TCNS), a secure technology solution that allows parties interested in constructing communications towers to ascertain whether their proposed construction would implicate a Tribal Nation's interest in culturally or religiously significant sites in the area of the proposed construction. TCNS is a non-public system that facilitates the notification, interactions, and review of proposed constructions, while protecting the private nature of both the particular applicant's proposed construction and the site of religious or cultural significance to a Tribal Nation. Use of TCNS immediately narrows the areas of interest of Tribal Nations to only those that they have indicated as relevant in this process and currently, every single Tribal Nation in the country has set its area of interest to participate in the TCNS system. A parallel system, E-106, exists to facilitate review of communications towers by interested state historic preservation officers (SHPOs).

On average the Commission processes more than 10,000 applications a year for new or collocated communications infrastructure through these systems. TCNS and E-106 provide an easy way to exchange information among Commission staff, Tribal Nations, SHPOs and applicants.

Facilitating Efficient PTC Infrastructure Deployment

In the spring of 2013, the railroads informed FCC staff that they intended to deploy PTC using a system that required the construction of approximately 22,000 wayside poles with antennas along the railroad tracks. Through continued discussions with the railroads we concluded, and the railroads have agreed, that the Commission's TCNS and E-106 processes, as they existed, were not suited to ensure efficient review of PTC wayside infrastructure on such a large scale. In fact, the existing TCNS and E-106 process was originally designed for tower-by-tower review and does not efficiently scale up to the large scale and linear nature of the proposed PTC projects. At the same time, we recognized that TCNS and E-106 do offer unique efficiencies that will provide many advantages to the railroads, FCC staff, SHPOs, and Tribal Nations. Consequently, the Commission immediately sought to develop alternative ways to leverage TCNS and E-106 so as to facilitate the deployment of necessary infrastructure while ensuring compliance with the relevant statutes and our rules. In doing so, we are mindful of the Commission's trust responsibility to consult with Tribal governments prior to implementing any regulatory action or policy that could significantly or uniquely affect Tribal governments, their land and resources.

Complicating our work to create an efficient process for new construction of PTC infrastructure is the fact that we must consider more than 10,000 PTC infrastructure deployments that were undertaken by the railroads without completing the environmental and historic preservation review clearly required by federal laws and regulations. This issue has figured prominently in our discussions with SHPOs and Tribal representatives, and in consultations with Tribal government officials, as we seek to create an efficient review process. Based on those discussions, it is clear to us that issues related to prior construction must be resolved contemporaneously with the issues surrounding future PTC infrastructure construction of up to an additional 20,000 wayside poles.

In May 2013, FCC senior staff advised the major freight railroads to defer submission of requests seeking SHPO and Tribal review of wayside pole deployments until the FCC could develop a more streamlined process specifically tailored to PTC wayside poles. FCC staff has consistently advised commuter railroads, which will deploy a limited number of wayside poles, to continue using the TCNS and E-106 systems.

THE PROGRAM COMMENT

Upon concluding that our existing processes were not suitable for PTC infrastructure review, FCC staff consulted with the Administration's Council on Environmental Quality (CEQ) and with the ACHP on the best path forward. CEQ confirmed that we could not simply exclude PTC wayside poles from the required NEPA review without a formal notice and comment rulemaking process to develop a categorical exclusion. ACHP similarly confirmed that there was no ready path to a broad exclusion from NHPA review. ACHP further advised that the most expeditious way for the Commission to streamline its historic preservation review process is to develop a "Program Comment."

A Program Comment is a document adopted and issued by ACHP that allows a federal agency such as the FCC to follow alternative historic preservation review processes that address specific needs of an agency. Under this approach, the FCC must develop a draft Program Comment and propose it to ACHP after consultation with all stakeholders, including government-to-government consultation with Tribal Nations. Adoption and implementation of the Program Comment does not require a formal FCC rulemaking.

We released a Section 106 scoping document — an initial step in developing a Program Comment – on September 27, 2013. We then met with Tribal officials and railroad representatives, in Rapid City, South Dakota in November 2013, and Tulsa, Oklahoma in December 2013. In an unprecedented fashion, in addition to face-to-face consultations between Federal and Tribal officials, these meetings also included sessions with all interested parties present. At their request, we also coordinated with the railroads to facilitate live demonstrations of their PTC deployments to Tribal officials.

As a result of those meetings and consultations, on January 8, 2014, the Commission advised the Class I freight railroads to begin submitting through TCNS and E-106 proposed PTC infrastructure along segments of track that were discussed during the Rapid City and Tulsa meetings using a Beta version of the process we envision implementing in the Program Comment. While the TCNS and E106 systems are not designed for public inspection, and thus do not have public view versions, we believe it is a good indication of forward progress that at least two of the Class I freight railroads have voluntarily filed information in TCNS, and four have filed forms in the E-106 system.

We began our PTC solutions dialogue with Indian Country in August of last year – the earliest practicable date - at the annual meeting of the National

Association of Tribal Historic Preservation Officers in Billings, Montana. We are scheduling additional consultations with Tribal Nations in the Pacific Northwest and Southwest to gather input on the draft Program Comment and continue the productive conversations between the Tribal Nations and the railroads. We also conducted a dialogue in early February with the Culture and Heritage Committee of the United South and Eastern Tribes, and we have conducted numerous other meetings with regional and national inter-Tribal organizations.

FCC staff has also engaged numerous SHPOs on the proposal to create a more efficient review process. In particular, we provided information to the SHPOs by letter in July 2013, October 2013, and January 2014, and we have conducted four teleconferences with the SHPOs in August 2013, December 2013, and February 2014.

The FCC publicly released a draft Program Comment on January 29, 2014. The draft Program Comment would apply to wayside infrastructure that will be constructed within existing railroad rights of way for use in the implementation and operation of PTC systems. Infrastructure eligible for the procedures of the Program Comment may support either wayside antennas or base stations, but must not: (1) be taller than 75 feet; or (2) be situated outside the railroad right-of-way. Eligible infrastructure also must create a foundation hole that is either (1) no greater than 20 inches in diameter (24 inches in the context of manual excavation) and no deeper than 15 feet; or (2) no greater than 30 square feet in surface area and no deeper than 6 feet. Applicants using the proposed Program Comment must ensure that associated cultural resource reports are prepared by professionals who meet the relevant standards in the Secretary of the Interior's Professional Qualifications Standards. Additionally, applicants must be willing to perform field surveys and/or support Tribal monitoring, where necessary to evaluate effects on historic properties, and to compensate Tribal preservation offices for professional services in accordance with ACHP's guidelines. The draft Program Comment seeks to promote administrative efficiency by providing for submission of poles in batches, rather than pole-by-pole, while ensuring that SHPOs and Tribal Nations will receive sufficient information about each pole to evaluate its effects on historic properties.

Comments on the draft Program Comment were due on February 12, 2014. The Commission submitted the Program Comment to ACHP on March 4, 2014. Under its rules, ACHP should act on the Program Comment by mid-April. Once adopted, the Commission is prepared immediately to begin accepting submissions from railroads under its terms.

COLLABORATION

Throughout this process, the Commission has been working closely with our partners at the Federal Railroad Administration, the National Transportation Safety Board, ACHP, CEQ, and the U.S. Fish and Wildlife Service to share information and ensure coordination between our agencies. As I stated we are also working closely with Tribal Nations and their cultural preservation officials, the SHPOs and, of course, the railroads subject to the PTC mandate. We will continue our close and productive collaboration with all of these parties to ensure that the Commission's processes do not inhibit the deployment of PTC by the statutory deadline.

In closing, let me reiterate that the Commission recognizes the importance of PTC to enhancing rail safety, and is committed to doing what is necessary to allow for an efficient and timely review process that ensures full consideration of the effects of PTC wayside poles on historic properties, including sites of religious and cultural significance to Tribal Nations, in a manner that allows the railroads to complete construction by the statutory deadline.

Thank you for the opportunity to appear today. I look forward to answering any questions you may have.

In: Transporting Crude Oil by U.S. Rail
Editor: Rosario S. McLaughlin
ISBN: 978-1-63117-837-5

Chapter 6

TESTIMONY OF EDWARD R. HAMBERGER, PRESIDENT & CEO, ASSOCIATION OF AMERICAN RAILROADS. HEARING ON "ENHANCING OUR RAIL SAFETY: CURRENT CHALLENGES FOR PASSENGER AND FREIGHT RAIL"*

On behalf of the members of the Association of American Railroads, thank you for the opportunity to discuss the current state of rail safety and specifically the transportation of crude oil by rail. AAR freight railroad members account for the vast majority of freight railroad mileage, employees, and traffic (including crude oil) in Canada, Mexico, and the United States.

The development of shale oil represents a tremendous opportunity for our nation to move closer to energy independence. The widespread benefits this would entail include reduced reliance on oil imports from unstable countries whose interests do not necessarily match up well with our own; increased economic development all over the country; thousands of new well- paying jobs; tens of billions in savings in our nation's trade deficit every year; and substantial amounts of new tax revenue for governments at all levels. Rail has a critical role in delivering these crucial benefits to our country.

---

* This is an edited, reformatted and augmented version of testimony presented March 6, 2014 before the Senate Committee on Commerce, Science, and Transportation, Subcommittee on Surface Transportation and Merchant Marine Infrastructure, Safety, and Security.

Let me make clear at the outset that, for our nation's freight railroads, pursuing safe operations is not an option, it's a business imperative. Most importantly, it's the right thing to do. Railroads are not just faceless corporations from somewhere far away. Rather, your neighbors are our neighbors. No matter where you live, chances are good that current or former rail industry employees live close by. We have an obligation to operate safely for their benefit and for the benefit of all members of the communities we serve.

Unfortunately, several recent serious rail accidents in Canada and the United States have led some to question railroads' ability to operate safely. We are committed to restoring their confidence and demonstrating that nothing is more important to railroads than the safety of their employees, their customers, and the communities in which they operate.

Railroads share the deep concern of members of this committee and the public at large regarding the safe transport of crude oil. From 2000 through 2013, a period during which U.S. railroads originated approximately 832,000 carloads of crude oil, more than 99.98 percent of those carloads arrived at their destination without a release caused by an accident. That said, railroads continue to look for ways to be safer. As the tragic accident last year in Quebec showed, and as reinforced by recent oil spills in North Dakota and Alabama, more work must be done to ensure public confidence in the transportation of crude oil by rail.

Working cooperatively with government agencies, our customers, our employees, and our suppliers, we're applying what we've learned over the past few years as rail crude oil traffic has surged to help ensure that our nation is able to safely and reliably utilize the tremendous national asset that domestic crude oil represents. This will be a true team effort involving shared responsibility among everyone involved in crude oil production, delivery, and consumption.

Railroads have long been doing their part — including taking actions that go beyond what legislation and regulations require — to maximize safety, and the industry will continue to take steps to further improve safety. As explained later in this testimony, these actions fall into three broad categories:

Accident Prevention

- Railroads are continuing to reinvest record amounts — their own funds, not taxpayer funds — back into their infrastructure and equipment. Despite a weak economy, railroads have invested far more back into their networks over the past five years — approximately $115 billion — than in any five-year period in history. This year, we project that railroads will invest more than $26 billion in their networks, more than ever before. One of the major aims of these investments is to make the rail network more robust, so that the industry's decades-long record of declining accident rates continues.

- Railroads have voluntarily agreed to institute speed restrictions, additional inspections, and other operational modifications for trains carrying large amounts of crude oil.
- By July 1, 2014, railroads will begin using a sophisticated statistical routing model that takes into account a variety of variables, including population density, rail traffic volume, track conditions, and availability of alternative tracks, so that trains with large amounts of crude oil can move on routes that pose the least overall safety and security risk.

Consequence Mitigation

- In 2011, the rail industry voluntarily adopted industry standards requiring new tank cars that carry crude oil to be built with additional safety features to reduce the probability of release should an accident occur. This year, railroads have called for even more robust new tank car standards for crude oil.
- Railroads have also called for an aggressive retrofit and phase out program for existing tank cars used to carry crude oil.

Emergency Response

- Railroads help communities develop and evaluate emergency response plans, and they train tens of thousands of emergency responders each year. The industry has agreed to spend several million dollars this year to develop an emergency response training program at the Transportation Technology Center in Pueblo, Colorado specifically geared to crude oil spills and to provide tuition assistance for some 1,500 emergency responders from across the country to attend that training.

Crude by Rail Meets Crude Oil Production Demand

The huge increase in rail crude oil volume is a function of the massive, salutary develop- ment of North American oil resources in recent years, especially “shale oil.” U.S. crude oil production peaked in 1970 at 9.6 million barrels per day, but by 2008 it had fallen to 5.0 million barrels per day as depletion of older fields outpaced new production. Over the past couple of years, however, technological advances in the extraction of shale oil, along with relatively high crude oil prices, have led to sharply higher U.S. crude oil production. The Energy Information Administration (EIA) states that production rose to an average of 6.5 million barrels per day in 2012 and 7.5 million barrels per day in 2013. Barring unforeseen circumstances, deposits of crude oil in shale formations across the country will continue to be developed. As a result, the EIA projects that U.S. crude oil production will increase to 8.5 million barrels per day in 2014 and 9.3 billion barrels per day in 2015.

Much of the recent increase in crude oil production has occurred in North Dakota, where crude oil production rose from an average of 81,000 barrels per day in 2003 to 940,000 barrels per day by the fall of 2013, making North Dakota the second-largest oil producing state.

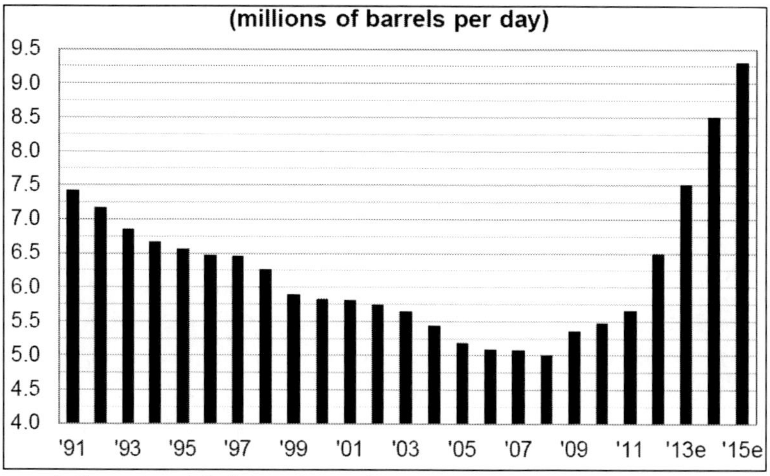

e - Energy Information Administration estimate.
Source: EIA.

U.S. Crude Oil Production.

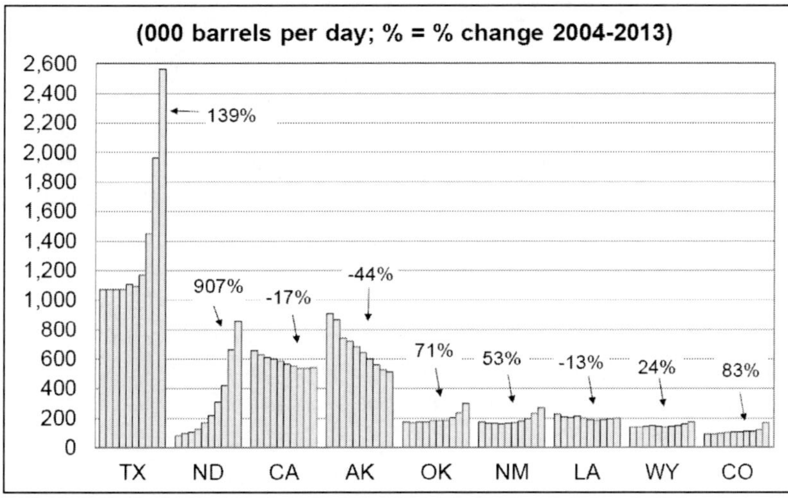

Source: EIA.

Avg. U.S. Crude Oil Production by State: 2004-2013.

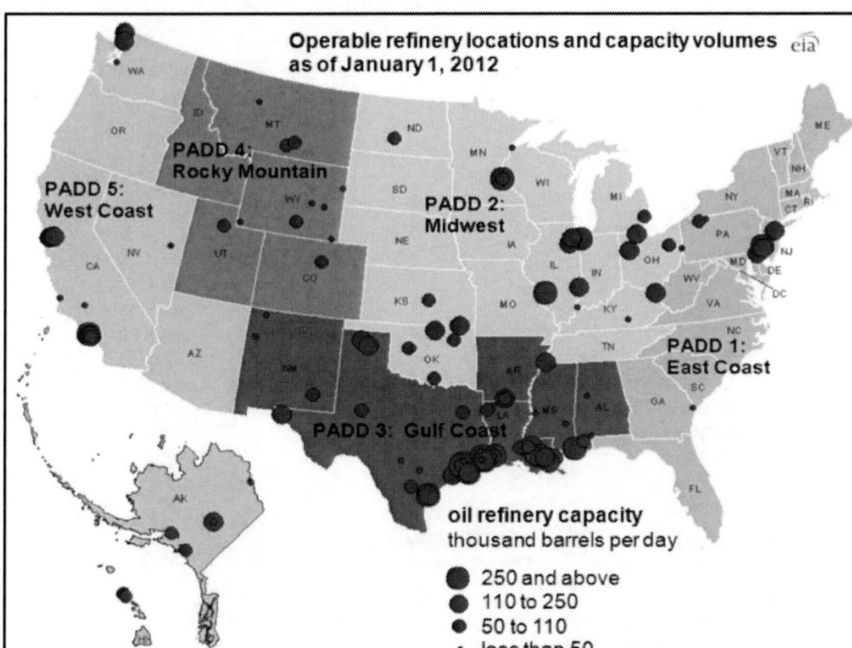

Of course, crude oil has little value unless it can be transported to refineries, but most U.S. refineries are located in traditional crude oil production areas (Texas, Oklahoma) or on the coasts where crude oil transported by tanker is readily accessible (California, Washington, New England, Gulf of Mexico). In part because of the long process required to obtain permits to build new refineries, it's unlikely that new refineries will come on line quickly near the new production areas.

Historically, most crude oil has moved from production areas to refineries by pipeline. However, in many of the new shale oil production areas, the existing pipeline network lacks the capacity to handle the higher volumes. Pipelines also lack the flexibility and geographic reach to serve many potential markets. Railroads, though, have the capacity and flexibility to fill this gap. In fact, the ability of a railroad to serve a refinery can make the difference between the refinery continuing to operate or closing down.

Railroads have seen dramatic recent increases in demand to transport crude. In 2008, U.S. Class I railroads originated 9,500 carloads of crude oil. By 2012, carloads had surged to nearly 234,000. Final numbers for 2013 aren't in yet, but we estimate that crude oil originations on Class I railroads in 2013 were around 408,000 carloads and terminations were around 434,000 carloads.[1] In 2013, crude oil accounted for about 1.4 percent of total originated carloads on Class I railroads, up from just 0.03 percent in 2008.

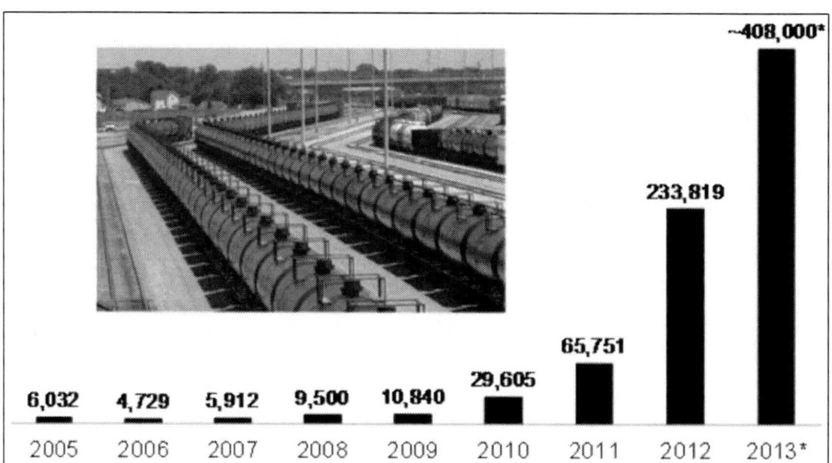

* Estimate based in preliminary data.
Source: AAR.

Originated Carloads of Crude Oil on U.S. Class | Railroads.

Assuming for simplicity that a rail tank car holds about 30,000 gallons (714 barrels) of crude oil, the approximately 408,000 carloads of crude oil originated by Class I railroads in 2013 equal around 800,000 barrels per day, or about 11 percent of U.S. crude oil production.

ADVANTAGES TO OUR NATION OF TRANSPORTING CRUDE OIL BY RAIL

Looking ahead, both pipelines and railroads will be needed to provide safe, reliable crude oil transportation for our nation. In addition to the critical fact that railroads provide transportation capacity in many areas where pipeline capacity is insufficient, railroads offer a number of other advantages for transporting crude oil:

- *Geographical Flexibility.* By serving almost every refinery in the United States and Canada, railroads offer market participants enormous flexibility to shift product quickly to different places in response to market needs.
- *Responsiveness.* Rail facilities can almost always be built or expanded much more quickly than pipelines and refineries can be. Essentially, railroads are the only transportation mode that can invest in facilities quickly enough to keep up with production growth in the emerging oil fields.
- *Efficiency.* As new rail facilities are developed, railroads are involved at every step, helping facility owners decide where to locate assets and how to lay out rail infrastructure to maximize safety and efficiency.
- *Underlying Infrastructure and Equipment.* Just over the past few years, railroads have invested tens of billions of dollars to replace and resurface tracks, buy new locomotives, build new terminals and track capacity, hire new employees, and take other steps to enhance their ability to transport crude oil.

Notwithstanding these attributes of rail, railroads recognize that if we are to continue down the path of energy independence, other transportation modes — including, of course, pipelines — have crucial roles to play.

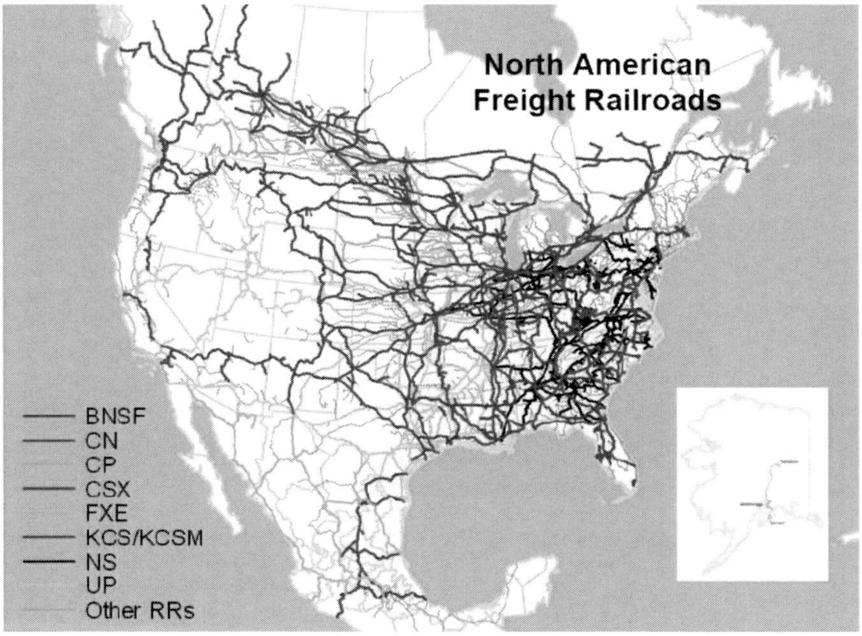

Working to Prevent Rail Accidents

We all know that, unfortunately, rail accidents happen, despite railroads' best efforts to prevent them, and some of those accidents have tragic consequences. Railroads take the challenge of moving the nation's crude oil extremely seriously, and they recognize that improving safety is an ongoing process.

The industry's commitment to safety is reflected in safety statistics from the Federal Railroad Administration (FRA). The three most common safety measures used by the FRA are train accident rates, employee injury rates, and grade crossing collision rates. From 2000 to 2013, the train accident rate fell 43 percent, with 2013 having the lowest rate ever, according to preliminary FRA data. The rail employee injury rate fell 48 percent from 2000 to 2013, with 2013 having the second-lowest rate ever (behind 2012). The grade crossing collision rate fell 42 percent from 2000 to 2013, with 2013 having the second-lowest rate ever (behind 2012).

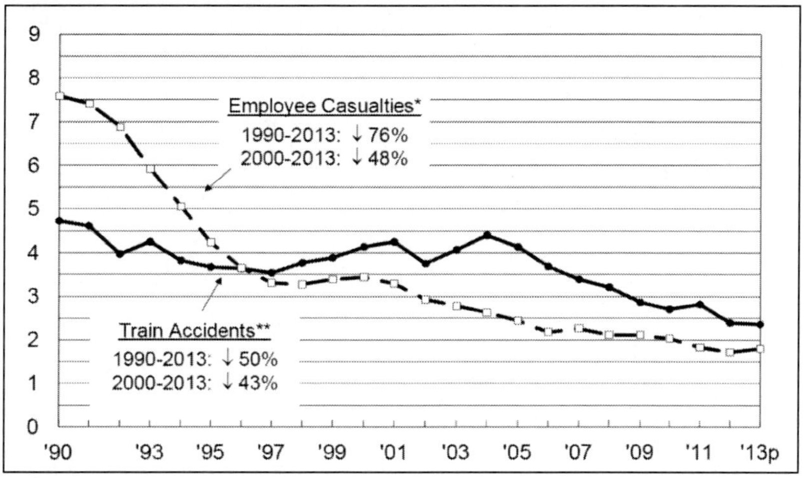

p – preliminary.

* Injuries & fatalities per 100 employee equivalents.

** Train accidents per million train-miles.

Source: Federal Railroad Administration.

U.S. Railroad Accident and Employee Injury Rates.

Collisions and derailments, two of the major categories that comprise the broader train accidents category, have fallen sharply over time as well. For example, according to FRA data, in 2000 there were 739 derailments involving freight trains on main lines or sidings, equivalent to 1.35 per million freight train miles.[2] In 2013, according to preliminary FRA data, there were 378 derailments (a 49 percent decline), equivalent to 0.70 per million freight train-miles (48 percent lower). In 2000, there were 45 collisions involving freight trains on main lines or sidings, equivalent to 0.08 per million train-miles. In 2013, the comparable figures were 28 collisions (down 38 percent) and 0.05 per million freight train-miles (down 37 percent).

Railroads are proud that their overall safety record, as measured by FRA data, has been trending in the right direction for decades. That said, it is an unfortunate reality that rail accidents occur, despite railroads' capital and labor intensive efforts to prevent them, and we know that some of those accidents have tragic consequences. Given the extent and complexity of rail operations — the railroad "factory floor" is outdoors and approximately 140,000 miles long — infrastructure and equipment sometimes fail and it is impossible to eliminate all risk of accidents. And while railroads respect and applaud the professionalism and attention to safety that rail employees bring every day to their jobs, human error continues to be a leading cause of accidents. Our goal is zero rail accidents, but when accidents do occur, we want to minimize and mitigate their negative impacts.

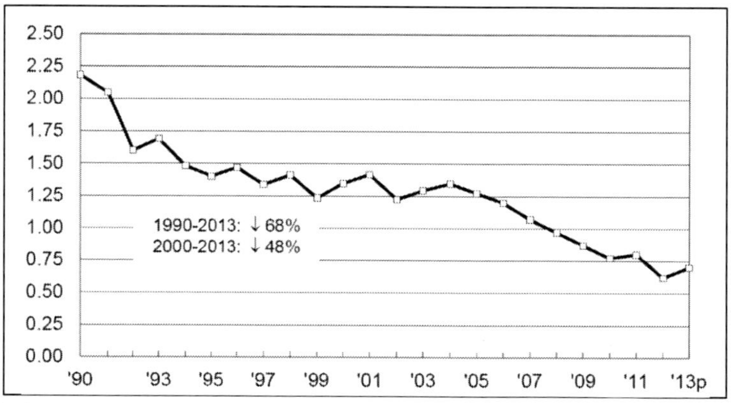

p - preliminary
* Per million freight train miles.
Source: Federal Railroad Administration.

Rates for Derailments Involving Freight Trains on Main Lines and Sidings.*

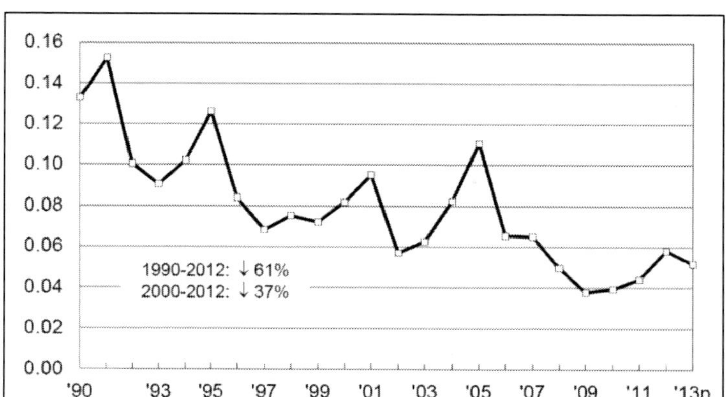

p – preliminary.
* Per million freight train miles.
Source: Federal Railroad Administration.

Rates for Collisions Involving Freight Trains on Main Lines and Sidings.*

One of the most important ways that railroads have reduced accidents is through significant and consistent investments back into their networks. In fact, in recent years, America's freight railroads have been reinvesting more private capital than ever before to renew, upgrade, and expand their infrastructure and equipment, including a record $25.5 billion in 2012 and a comparable amount in 2013. Rail spending this year is expected to be even higher. The vast majority of these investments have improved rail safety directly or indirectly. In fact, for many of these investments, improving safety is the primary reason the investments are made.

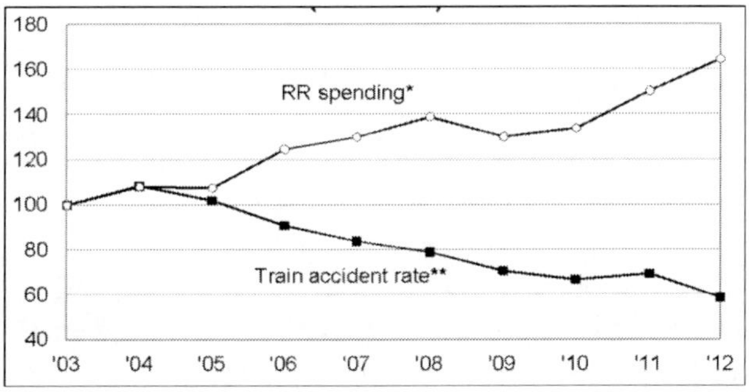

* Capital spending + maintenance expenses on infrastructure and equipment.
** Total train accidents per million train-miles.
Source: AAR, FRA.

Rail Spending on Infrastructure and Equipment* vs. Overall Train Accident Rate** (2003 = 100).

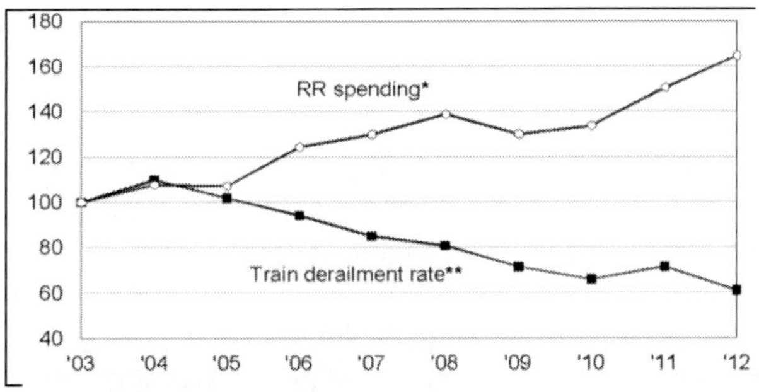

* Capital spending + maintenance expenses on infrastructure and equipment.
** Derailments per million train-miles.
Source: AAR, FRA.

Rail Spending on Infrastructure and Equipment* vs. Train Derailment Rate** (2003 = 100).

Just from 2008 to 2012, Class I railroads spent nearly $26 billion in capital expenditures on new crossties (77 million), new rail (2.9 million tons), and new ballast (nearly 61 million cubic yards). Over the same period, they spent billions of additional capital expenditure dollars on signal and communications systems, bridges and tunnels, and machinery. These and other capital investments, as well as the billions of dollars railroads spend on maintenance of their networks each year, have made railroads much safer. And as the charts above show, there is a clear correlation between rail reinvestments and rail safety improvements.

The Role of Safety-Enhancing Technologies

At a basic level, railroading today seems similar to railroading 150 years ago: it still consists of steel wheels traveling on steel rails. This apparent similarity, however, masks a widespread application of modern technology and a huge variety of ongoing initiatives to research, test, and apply advanced technologies to yield the safety record of continuous improvement experienced by the railroad industry.

Many of these advancements were developed or refined at the Transportation Technology Center, Inc. (TTCI), the finest rail research facility in the world, in Pueblo, Colorado. TTCI is a wholly owned subsidiary of the AAR. Forty-eight miles of test tracks, highly sophisticated testing equipment, metallurgy labs, simulators, and other diagnostic tools are used to test track structure, evaluate freight car and locomotive performance, assess component reliability, and much more. The facility is owned by the FRA but has been operated (under a competitively bid contract with the FRA) by TTCI since 1984.

A few of the many examples of safety-enhancing rail technologies developed at TTCI or elsewhere that have come into widespread use or are in the process of being implemented include:

- *Wayside detectors* that identify defects on passing rail cars, including overheated bearings and damaged wheels, dragging hoses, deteriorating bearings, cracked wheels, and excessively high and wide loads.
- *Internal rail inspection systems* using induction or ultrasonic technology to detect internal flaws in rails which are caused by fatigue and impurities introduced during manufacturing.

- *Track geometry vehicles* that use electronic and optical instruments to inspect track alignment, gauge, curvature, and other track conditions. Processed data from track geometry cars can help railroads determine when track needs maintenance.
- New automated detector systems are being tested and evaluated by TTCI to inspect the under carriage, safety appliances and freight car components using *machine-vision-based car inspection systems*.
- *Ground-penetrating radar* that helps identify problems (such as excessive water penetration and deteriorated ballast) that hinder track stability.
- Because a relatively small share of freight cars cause an inordinately high percentage of track damage and have a higher than usual propensity to derail, TTCI is working on ways to use *optical geometry detectors* to identify poorly performing freight car trucks.[3]
- *Positive train control (PTC)* systems, designed to automatically stop or slow a train before certain accidents occur, are being developed and implemented. PTC is discussed more fully later in this testimony.

Railroads and their suppliers will continue to pursue these and other technological advances that make rail transportation safer and more secure.

Rail industry safety will also be enhanced by the Asset Health Strategic Initiative (AHSI), a multi-year rail industry program that will apply information technology processes to improve the safety and performance of freight cars and locomotives across North America. In a nutshell, AHSI aims to improve safety and reduce costs across the rail industry by addressing mechanical service interruptions, inspection quality, and yard and shop efficiency. It is based on the recognition that improving asset health means more than just focusing on railcar and locomotive repair. Rather, it encompasses the entire rolling stock health cycle, incorporating prevention, detection, planning, movement, and repair.

MODIFYING RAIL INDUSTRY PROCEDURES TO ENHANCE CRUDE OIL SAFETY

To enhance the safety of crude oil movements, railroads are voluntarily making operational modifications for trains carrying large amounts of crude oil. These modifications are the result of consultations with Secretary Foxx and other officials at the U.S. Department of Transportation (DOT), Administrator Szabo at FRA, Administrator Quarterman at PHMSA, and other government policymakers. Railroads share their vision for making a safe rail network even safer, and we're pleased that we were able to work together to pinpoint new operating practices that enhance the safety of moving crude oil by rail.

First, *routing*. Several years ago, the FRA, PHMSA, the Transportation Security Administration, the Federal Emergency Management Agency, and the railroads jointly developed the Rail Corridor Risk Management System (RCRMS), a sophisticated statistical routing model designed to aid railroads in analyzing and identifying the overall safest and most secure routes for transporting TIH materials. The model uses a minimum of 27 risk factors —

including hazmat volume, trip length, population density along the route, availability of alternate routes, and emergency response capability — to assess the overall safety and security of rail routes. The FRA regularly audits railroads' use of the RCRMS. Major railroads have agreed that, no later than July 1, 2014, they will begin using the RCRMS for trains carrying at least 20 carloads of crude oil.

Second, *speed restrictions*. Back in August 2013, railroads self-imposed a 50-mph speed limit for trains carrying 20 or more carloads of crude oil. Beginning no later than July 1, 2014, if a train is carrying at least 20 cars of crude oil and at least one of those cars is an older "DOT-111" car (these cars are discussed further in the next section), that train will travel no faster than

40 mph when travelling within one of the 46 nationwide "high threat urban areas" designated by the Department of Homeland Security.[4] In addition, railroads will continue to work with communities through which crude oil trains move to address, on a location-specific basis, concerns that the communities may have.

Third, *inspections*. Comprehensive FRA regulations dictate the various kinds of inspections railroads are required to perform and how often. A description of the full range of inspections that railroads undertake is beyond the scope of this testimony, but suffice it to say that the FRA-mandated inspection regime is comprehensive and thorough.

New FRA regulations regarding inspections for internal rail defects will become effective on March 25th. Railroads have agreed that, going forward, for main line tracks on which trains carrying at least 20 carloads of crude oil travel, they will perform at least one more internal rail inspection each calendar year than the new FRA regulations require. In addition, railroads will conduct at least two automated comprehensive track geometry inspections each year on main line routes over which trains with 20 or more loaded cars of crude oil are moving.[5] The FRA regulations do not require railroads to perform automated comprehensive track geometry inspections.

Fourth, *defect detectors*. Railroads will make sure that specialized track side "hot box" detectors are installed at least every 40 miles along routes with trains carrying 20 or more cars containing crude oil.[6] These detectors help prevent accidents by measuring if wheel bearings are generating excessive heat, which is a warning that the bearings are in the process of failing.

Fifth, *braking*. By April 1, 2014, railroads will make sure that trains operating on main line tracks carrying at least 20 carloads of crude oil are equipped either with distributed power locomotives (i.e., locomotives placed in locations other than the front of the train) or with two- way telemetry end-of-train devices. These technologies allow train crews to apply emergency brakes simultaneously from both the head end and locations further back in the train in order to stop the train faster.

Mitigation Through Enhancing Tank Car Safety

While railroads pursue continuous improvement in safe operations, including most recently the new voluntary actions described above, it's important also for tank car standards to evolve to mitigate the consequences of an incident when one occurs. Crude oil and thousands of other products are transported in tank cars. The total North American tank car fleet consists of about 335,000 cars. Railroads themselves own less than 1 percent of these cars; nearly all are owned by rail customers and leasing companies. The dozens of distinct types of tank cars are differentiated by characteristics (pressure or general service, insulated or non-insulated, how much they can carry, and so on) that make them suitable or not suitable for carrying particular commodities. Approximately 228,000 tank cars are so-called "DOT-111" general service tank cars. These cars are considered the workhorses of the tank car fleet. According to the most recent figures, around 98,000 DOT-111 cars are used to transport crude oil or other flammable liquids.

In the United States, federal regulations pertaining to tank cars are set by PHMSA. Transport Canada performs a similar role in Canada. In addition, the AAR Tank Car Committee sets industry standards regarding how tank cars used in North America are designed and constructed. These standards are often above and beyond federal standards. The Tank Car Committee is comprised of railroads, rail car owners, rail car manufacturers, and rail hazmat customers, with active participation from the U.S. DOT, Transport Canada, and the National Transportation Safety Board (NTSB).

The rail industry has been aggressively searching for ways to improve tank car safety. For example, in March 2011, the AAR petitioned PHMSA to adopt more stringent requirements for new tank cars used to transport certain types of hazardous materials, including crude oil. These tougher standards called for more puncture resistance through the use of a thicker tank car shell or a jacket, extra protective half-height (at a minimum) "head shields" at both ends of tank cars, and additional protection for the fittings on the top of a car that enable access to the inside of the car.

In July 2011, after it had become clear that PHMSA approval of the AAR's proposal was not imminent, the Tank Car Committee adopted what the AAR had proposed to PHMSA as the basis for new industry standards for tank cars used to carry ethanol or crude oil. The new standards, referred to as

"CPC-1232," apply to new tank cars ordered after October 1, 2011. To date, around 18,000 tanks cars have been built to this tougher CPC-1232 standard.

More recently, in November 2013, the rail industry called on PHMSA to adopt standards even more stringent than CPC-1232 for new tank cars used to transport crude oil and ethanol. The November 2013 proposal also called for aggressively retrofitting or phasing out of tank cars (including CPC-1232 cars and the older DOT-111 cars) used to transport crude oil or ethanol. The November 2013 proposal recognizes that input is needed from shippers and tank car manufacturers to determine the precise parameters of a phase-out program and to identify the retrofits that should be required.

Since the November 2013 proposal, the rail industry has continued to evaluate what other standards might be appropriate to make tank cars used to transport certain commodities, including crude oil, more robust. For example, railroads now support strengthening tank cars used to transport crude oil with thicker, 9/16th inch shells.[7] Railroads also would require that tank cars be equipped with jackets and thermal protection, as well as full-height head shields, top fittings protections, and bottom outlet handles that will not open in a derailment.

The graphic on the next page summarizes proposed rail industry standards for tank cars carrying crude oil.

Through these additional standards and other means, railroads are continuing to work with other stakeholders to enhance rail safety and provide certainty to all stakeholders. These efforts reflect the rail industry's long-standing belief that the safety of crude oil by rail is a shared responsibility among all stakeholders in the crude oil supply chain.

The concept of shared responsibility also applies to accurate and timely determinations as to the chemical characteristics of the crude oil railroads are asked to transport. Under federal regulations, the entity "offering" the crude oil to the railroad for transport (*e.g.*, the oil producer) is responsible for properly classifying the oil based on its level of hazard. On February 25, the FRA issued an executive order requiring that crude oil from the Bakken region be tested to ensure that it is properly classified before it is transported by rail. Railroads support the pursuit of proper classification and labeling of petroleum crude oil in tank cars by shippers prior to transport. This is essential to ensuring that first responders are able to safely and appropriately respond in the event of an accident or incident.

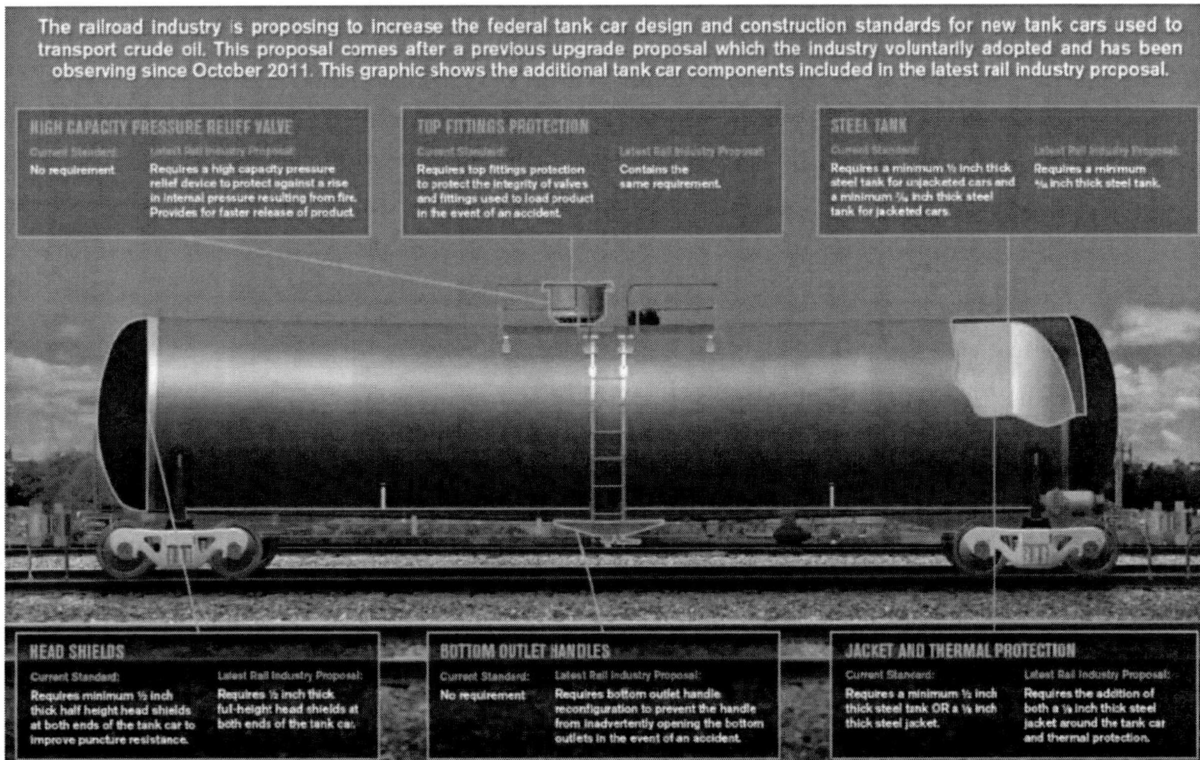

Source: Association of American Railroads, February 2014.

Evolution of Rail Industry Tank car standards for crude oil.

IMPROVING EMERGENCY RESPONSE: PARTNERSHIPS WITH FIRST RESPONDERS AND SHIPPERS

The railroads have extensive emergency response functions, which work in cooperation with federal, state and local governments, especially since 9/11. Sharing important information about shipments is part of that. Upon request, railroads provide appropriate local authorities with a list of the hazardous materials, including crude oil, transported through their communities. It is simply not true, as one sometimes hears, that railroads refuse to provide this type of information.

Significantly, more than 25 years ago, the AAR established what is now the Security and Emergency Response Training Center (SERTC), a world-class facility in Pueblo, Colorado, that is operated by TTCI. The SERTC has provided in-depth hazmat emergency response training to more than 50,000 emergency responders and railroad and chemical industry employees.

In 2007, Congress authorized the National Domestic Preparedness Consortium (NDPC), a consortium within the Department of Homeland Security and funded by FEMA. The purpose of the NDPC is to identify, develop, test, and deliver training to the nation's emergency first responder community. Of the NDPC's seven members, only one — SERTC — is specifically designed to provide first responder training for rail and other surface transportation accidents. No other training center in the country possesses comparable infrastructure, including dozens of freight and passenger railcars, highway cargo tanks, intermodal containers, van trailer, and even a barge. Live simulations deliver tactical intervention training with unsurpassed realism.

Unfortunately, since it was added to the NDPC in 2007, SERTC has received only $10 million from FEMA for surface transportation first responder training — $5 million in 2009 and $5 million in 2010. We understand that budgets throughout the federal government have been challenged in recent years. That said, Congress provided an additional $5 million to the NDPC in the 2014 omnibus appropriations act, increasing NDPC funding from $93 million to $98 million. FEMA has yet to allocate those dollars among the NDPC members. We hope this committee would agree that allocating these funds to enhance the ability of first responders to respond to crude oil incidents would be a sensible step for FEMA to take.

In addition to SERTC, as part of their regular operations, railroads and communities develop and evaluate emergency response plans and train more than 20,000 emergency responders each year. Through their own efforts and working in conjunction with the Transportation Community Awareness and Emergency Response Program (TRANSCAER), railroads will continue to work closely with emergency responders in the communities they serve so that damage caused by train accidents can be minimized.

In addition to their other ongoing emergency response training efforts, the AAR recently announced that railroads will provide approximately $5 million by July 1 of this year to develop a specialized crude-by-rail training and tuition assistance program for local first responders. The funds will be used to design a curriculum at TTCI specifically devoted to crude oil emergency response, to provide tuition assistance for an estimated 1,500 first responders to attend TTCI for training, and to provide additional training to local emergency responders closer to home.

By July 1, 2014, railroads will also develop an inventory of resources for emergency responders along routes over which trains with 20 or more cars of crude oil operate. This inventory will include locations for the staging of emergency response equipment and contacts for the notification of communities. When the inventory is completed, railroads will provide the DOT with information on the deployment of the resources and will make the information available upon request to appropriate emergency responders.

Finally, it is sometimes claimed that railroads bear no costs for cleanup of oil spills and that the entire response burden falls on local responders. That's not true. Emergency responders have control of railroad accidents in which crude oil (or any other hazardous material) is spilled, but railroads often provide the resources for mitigating the accident. As noted above, railroads have emergency response plans in place to mobilize the labor and equipment necessary to mitigate accidents. Railroads also reimburse local emergency agencies for the costs of materials the agencies expend in their response efforts.

ADDRESS SAFETY CONCERNS, BUT DON'T LOSE SIGHT OF THE BENEFITS OF DOMESTIC CRUDE OIL

The development of crude oil resources in recent years represents a tremendous opportunity for this country, including the opportunity to move toward energy independence. As a report earlier this year from the Congressional Research Service (CRS) notes, "the prospect of U.S. energy independence is grounded in the production growth from tight oil formations such as the Bakken Formation in North Dakota and Montana, and the Eagle Ford Formation in Texas."[8] CRS says that, "Relative to other fuels, the United States is more dependent upon imports for its oil requirements, still accounting for almost 40% of consumption," but "since June 2005, when crude oil imports reached a peak, they have dropped almost 3.3 million [barrels per day], or 23%, through October 2013."

It's difficult to overstate the economic and security benefits associated with continued growth in domestic crude oil production. Over time, it will mean reductions in the nation's trade deficit of tens of billions of dollars every year. It will mean new and better employment opportunities for hundreds of thousands of Americans and better economic development opportunities for regions all over the country. It will mean billions of dollars in new tax revenues for governments at all levels. And it will mean less reliance on sources of oil from places in the world that are not secure and whose interests do not necessarily correspond well to those of the United States. The Peterson Institute, a well-respected, nonprofit, and nonpartisan research institution devoted to the study of international economic policy, recently found that, along with lower energy costs, the growth in domestic energy production should increase annual U.S. GDP growth between 0.09 and 0.19 percentage points through 2020. That adds up to hundreds of billions of dollars in higher GDP.

As we discuss ways to enhance the safety of transporting crude oil by rail, these important benefits should be kept in mind.

Positive Train Control

The Rail Safety Improvement Act of 2008 (RSIA) requires Class I freight railroads to install a fully functioning, nationwide positive train control (PTC) network by the end of 2015 on main lines used to transport passengers or TIH materials. Specifically, PTC as mandated by Congress must be designed to prevent train-to-train collisions; derailments caused by excessive speed; unauthorized incursions by trains onto sections of track where maintenance activities are taking place; and the movement of a train through a track switch left in the wrong position. The technology must also be fully interoperable, meaning that the system in place on any one railroad must be able to seamlessly interface with the system on any other railroad.

PTC is an unprecedented challenge, both in terms of the technologies to be used and the integration of those technologies. A properly functioning PTC system must be able to determine the precise location, direction, and speed of trains; warn train operators of potential problems; and take immediate action if the operator does not respond to the warning provided by the PTC system. For example, if a train operator fails to begin stopping a train before a stop signal, the PTC system would apply the brakes automatically before the train passed the stop signal.

Railroads have been devoting massive resources to PTC. They've retained more than 2,200 signal system personnel to implement PTC, and to date have spent approximately $4 billion (of their own funds, not public funds) on PTC development and deployment. They expect to spend that much again — approximately $8 billion in total — before development and installation is complete. Hundreds of millions of dollars will be spent each year after that to maintain the system.

PTC's complexity, the enormity of the implementation task, and the fact that much of the technology and engineering applications PTC requires have had to be developed from scratch mean that, despite railroads' best efforts, much work remains to be done. The many potential failure points in PTC systems must be identified, isolated, and corrected, and the system must be made fully interoperable across all of the nation's major railroads — all without negatively affecting existing rail operations.

Railroads also face serious non-technological barriers to timely PTC implementation. Today, the most serious such challenge involves PTC antenna structures.

At its heart, PTC is a massive communications system. Locomotives must be able to communicate with the "back office" concerning the train's speed, location, and many other parameters and receive information regarding, among many other things, the locations of other trains in the area, possible schedule changes, safety alerts, and so on.

This back-and-forth communication can take place only if a sophisticated, comprehensive wireless communications network is in place. A key part of this network is a series of thousands of antennas, spaced (on average) every few miles along the 60,000 or so miles over which PTC is being installed on U.S. freight railroads. These antennas are generally around 40-feet tall, and the vast majority are to be installed directly adjacent to the tracks on existing railroad rights-of- way, owned by the railroads themselves, in holes just a few feet deep and a couple of feet wide. In total, approximately 22,000 PTC-related antennas need to be installed.

The railroad industry began working several years ago with the Federal Communications Commission (FCC) to license the wireless spectrum necessary for PTC, and to its credit, the FCC has worked diligently to address spectrum-related issues. Nonetheless, the industry learned just last year that, under the FCC's interpretation of Section 106 of the National Historic Preservation Act (NHPA), railroads must ascertain, on an antenna-by-antenna basis, if the antennas will negatively impact areas of historic, cultural, or religious significance.

At the center of the challenge is the FCC's required notification and evaluation process that utilizes the FCC's "Tower Construction Notification System" (TCNS) for review by Native American tribes. Under that system, railroads must input certain information into the TCNS. That information is then transmitted to any Native American tribe that has expressed interest in the county in which an antenna will be located.

The initial information that railroads must input into the TCNS, such as the precise location of the antenna to be installed and its height, is relatively straightforward. That information is then transmitted to interested tribes through various means, and the tribe has up to a couple of months to expresses interest in the proposed antenna site. If it does so, it can demand much more comprehensive information about that site — such as a complete archaeological history — that can be difficult (if not impossible), costly, and time consuming for the railroad to obtain. Based on railroads' experiences to date, it takes, on average, three to five months between the time the railroad initially inputs information into the TCNS and the time when all necessary reviews are completed and the antenna can be cleared for installation. Multiply this process by 22,000 antennas and it becomes very clear why this is such a significant issue.

As stated earlier, the vast majority of PTC antennas are not large and are to be installed in small holes on railroads' own rights-of-way. Many of the rail lines in question have been in use for decades, often for well over 100 years. Generally speaking, the rights-of-way at issue have been disturbed countless times in the past as railroads performed standard maintenance, installed other types of signal systems, built culverts, improved drainage, or undertook any of innumerable other activities related to rail operations and infrastructure construction and upkeep.

An example of a PTC antenna structure near other railroad signals.

Once railroads started using the TCNS last year, it quickly became clear that the system was woefully inadequate for a deployment on the scale of PTC and in the time frame mandated by the RSIA. In fact, shortly after railroads began using the TCNS, the FCC asked them to stop using it while the agency developed a new process for PTC antennas. That was around ten months ago.

After nearly a year of discussion among various parties, during which the installation of wayside antennas ceased, on January 29, 2014, the FCC proposed what it calls a "streamlined" process for PTC-related reviews. Unfortunately, we do not believe the "streamlined" process will lead to a meaningful reduction in the substantial and excessive delays associated with PTC antenna installation.

Under the streamlined system, the FCC would still require an antenna-by-antenna evaluation. Perhaps most vexing, even when a tribe cannot identify any specific historic or cultural area or property that could potentially be impacted at any proposed individual antenna site, the tribe can still demand a comprehensive review of the site, which could include field work and the preparation of wide-ranging cultural resource reports and ethnographic studies.

The tribe can also demand that railroads dig holes for antenna structures by hand. These demands, which the streamlined process allows, extend well beyond what is required under the NHPA. Moreover, the streamlined process does not establish firm deadlines by which the FCC will resolve disputes regarding sites, and it is not clear the agency has the resources to manage disputes around potentially thousands of antenna sites.

Included as an appendix to this testimony is a copy of the AAR's recent comments on the FCC's streamlined proposal. As the AAR comments make clear, the FCC has ample authority to exempt all PTC-related infrastructure no taller than 75 feet located on the railroad right-of-way and not immediately proximate to a known historic property. We respectfully suggest that these antennas should be exempted. If the FCC decides not to pursue an exemption, it should put in place a process that really does expedite the historic review process, provides deadlines on the resolution of disputes, and encompasses all railroads' PTC infrastructure on the right-of-way. The existing "streamlined" process does none of these things.

Just to be clear, railroads are not seeking exemptions regarding PTC antenna installations on Native American reservations or similar areas, or in areas that are shown to be of significant historical or cultural interest. In these cases, railroads are happy to work with state preservation officials and tribes. Rather, railroads respectfully suggest that the NHPA was not designed and should not be used to impede PTC antenna installation in the way it has come to.

The bottom line is that without further changes to the FCC approval process, the timeline for ultimate deployment of PTC will be delayed significantly. The 2013 construction season was lost for PTC wayside antennas. A new review process at the FCC will not be in place until at least April of this year. If that process takes several months to clear locations, the 2014 construction season will also be in jeopardy. Railroads will continue to work with the FRA and the FCC in good faith to try to find a workable solution to this issue. Please keep in mind, though, that these antennas are at the heart of the PTC network, and there is no realistic "work around." Until railroads are able to install these antennas in large numbers, huge portions of the PTC testing and implementation process simply cannot proceed.

Railroads have been working extremely hard to meet the 2015 PTC deadline, and they will continue to aggressively pursue PTC implementation. However, due to the significant technological challenges associated with PTC development and installation, the eight-month construction moratorium imposed by the FCC, and the remaining PTC regulatory uncertainty, railroads believe that the existing PTC implementation deadline of December 31, 2015 will need to be extended. Doing so would allow railroads, the FRA, and others to make sure PTC is done right.[9] In the meantime, incremental PTC implementation would continue, meaning that more and more of the safety benefits of PTC would be coming on line.

Conclusion

Our nation's freight railroads share the urgency of this committee and the public at large regarding the need to augment the safe transport of products by rail. Railroads will continue to work with the Administration, their customers, and other stakeholders as necessary to identify additional safety enhancing steps that will make the North American rail network even safer.

Appendix

Before the Federal Communications Commission Washington, DC 20554

In the Matter of	)	
	)	
Comment Sought on Draft Program Comment to	)	WT Docket
Govern Review of Positive Train Control	)	No. 13-240
Facilities under Section 106 of the National	)	
Historic Preservation Act	)	

Comments of the Association of American Railroads

Louis P. Warchot
Senior Vice President-Law and General Counsel

Timothy J. Strafford
Assistant General Counsel
The Association of American Railroads
425 3rd Street, S.W. Suite 1000
Washington, DC 20024

Michele C. Farquhar
Deborah K. Broderson
Hogan Lovells US LLP
Counsel to the Association of American Railroads
555 13th Street, NW
Washington, DC 20004

(202) 639-2502 (202) 637-5663
February 12, 2014

Executive Summary

Congress has mandated that a nationwide Positive Train Control ("PTC") network be fully operational on the nation's passenger and freight railroads by December 31, 2015. As part of satisfying this statutory mandate, the railroads must install approximately 22,000 wayside poles and other PTC-related infrastructure on the national railroad rights of way. However, nearly ten months after the Federal Communications Commission ("FCC" or "Commission") promised a solution to expedite the historic preservation review of these facilities—during which time all review was suspended and the entire 2013 construction season lost—the draft Program Comment offers little that would permit the timely deployment of these poles. The Association of American Railroads ("AAR") continues to feel strongly that the best solution to address the need for expedited, comprehensive historic preservation review of PTC wayside facilities remains an exemption from review for all PTC-related infrastructure no taller than seventy-five feet located on the railroad rights of way and not immediately proximate to a known historic property.

If the FCC declines to seek such an exemption, the Commission should adopt a Program Comment that truly expedites the historic preservation review process for such facilities, imposes firm deadlines on the resolution of the approval process, and ensures that all railroads can benefit from these revisions—which the current draft Program Comment does not accomplish. Specifically, although the draft Program Comment provides for a conditional exemption for State Historic Preservation Officer review, it:

- Will not apply to any PTC-related infrastructure deployed by at least three of the seven Class I railroads;
- Continues to anticipate time-consuming, pole-by-pole Tribal review;
- Does not offer definitive deadlines for the resolution of review by Tribal Nations;
- Allows Tribes to make automated information requests which have the effect of delaying approval;

- Shifts the burden, properly assigned to Tribes, to the railroads to identify historic properties of cultural and religious significance to Tribal Nations;
- Requires the time-consuming and costly preparation of cultural resource reports that would not be helpful to many Tribal Nations, and that are not required by the existing historic preservation review process; and
- Allows Tribal Nations to require monitoring and/or alternative excavation techniques for potentially every PTC-related deployment site nationwide with no evidentiary showing of the likelihood of the presence of cultural resources.

As provided in the rules of the Advisory Council on Historic Preservation ("ACHP"), program comments are intended to preclude case-by-case review of undertakings. In contrast, the draft Program Comment anticipates the separate review by Tribal Nations of each of the 22,000 PTC-related wayside facilities that must be deployed by the railroads pursuant to a Congressional mandate. Moreover, the proposed historic review process does not provide firm deadlines for the resolution of Tribal consultation, which in the experience of the railroads can take as long as nine months—*for each pole*. The process outlined in the draft Program Comment would not only foreclose the industry from meeting the PTC implementation deadline imposed by Congress, but almost certainly lengthen the deployment process well beyond 2015.

The FCC has the authority to establish firm deadlines for Tribal Nation review, and should do so. Specifically, agencies need only provide a Tribal Nation a "reasonable opportunity" to identify concerns regarding historic properties, and other agencies have imposed decisive deadlines on matters of Tribal consultation. The AAR appreciates the intent behind the conditional exemptions from State Historic Preservation Officer review that the FCC has included in the draft Program Comment, but remains concerned that, as drafted, the Program Comment does not represent a significant improvement over the Commission's existing Tribal review process, and in some ways is more onerous. Neither process is suitable for the review of a critical, time-sensitive public safety initiative with the broad, national scope of PTC.

The ACHP's rules do not require that the historic preservation review process be exhaustive, but simply that Federal agencies make a "good faith and reasonable" effort to identify historic properties and "take into account" the effects of their undertakings on such properties. The PTC-related wayside facilities that must be deployed are not 300 foot towers being dug into undisturbed land, but small poles closer in height to standard utility poles already ubiquitous in the urban and rural landscape, and located on the railroad rights of way in soil that has been subject to repeated disruption for, in some cases, well over a hundred years.

Importantly, facilities located on the railroad rights of way that have previously been put through the existing historic preservation review process were ultimately found to have no effects on any historic properties, and there is no reason to believe that any of the remaining PTC-related wayside installations will be any different. The draft Program Comment wrongly turns the historic review process on its head and creates a presumption that every site proposed for PTC deployment should be considered a historic property unless shown otherwise by the railroads. In fact, the burden lies on Tribal Nations to establish, through evidence supporting a high probability of the presence of archeological artifacts, that a historic property exists that should be the subject of consultations and, if needed, mitigation.

Finally, while the railroads are prepared to work closely with the FCC, Tribal Nations, State Historic Preservation Officers, and other stakeholders to address concerns regarding previously constructed PTC facilities, resolution of this issue should not be a precondition for arriving at a workable Section 106 process that permits the timely installation of the thousands of remaining wayside structures required for nationwide PTC deployment. The focus of the FCC, and the draft Program Comment, should be on fulfilling its obligation under the ACHP's rules and enabling the expedited, programmatic review of all PTC-related facilities located on the railroad rights of way.

I. Introduction

The Association of American Railroads ("AAR")[10] respectfully submits these comments in response to the Public Notice ("*Public Notice*") released by the Wireless Telecommunications Bureau of the Federal Communications Commission ("FCC" or "Commission") in the above- captioned proceeding.[11] In the *Public Notice*, the FCC seeks comment on the draft Program Comment[12] to govern review of Positive Train Control ("PTC") wayside facilities under Section 106 of the National Historic Preservation Act ("NHPA").[13]

The AAR continues to feel strongly that the best solution to address the need for expedited, comprehensive review of PTC wayside facilities remains an exemption from State Historic Preservation Officer ("SHPO") and Tribal review for all PTC-related wayside facilities no taller than seventy-five feet located on the railroad rights of way that are not immediately adjacent to any previously recorded historic properties, as detailed in the AAR's previous comments on the PTC Scoping Document.[14] In the alternative, the Commission should modify the draft Program Comment to truly expedite the Section 106 review process, and ensure that all railroads can benefit from these modifications.[15] As currently drafted, the Program Comment provides beneficial exemptions regarding SHPO review, but fails to significantly address the existing deficiencies in the FCC's Section 106 Tribal review process. In particular, a request for additional information by a Tribal consulting party can eliminate all deadlines for the resolution of review. Moreover, the structure of the draft Program Comment inverts the normal Section 106 process by requiring that the railroads supply consulting Tribal Nations with information regarding historic properties of potential cultural and religious significance to those Tribes. Neither the Commission's existing historic preservation review mechanism, nor the process outlined in the draft Program Comment, is suitable for the review of a critical, time-sensitive public safety initiative with the broad, national scope of PTC.

II. The Draft Program Comment Will Not Expedite the Current Section 106 Review Process

After waiting nearly ten months for the release of a streamlined solution to the existing historic review process from the FCC, the railroads are disappointed that the draft Program Comment offers no appreciable improvements regarding the Tribal review process. Instead, like the existing Section 106 review mechanism, the draft Program Comment is poorly suited to expedite review of large scale time sensitive infrastructure projects such as PTC, and in particular lacks clear deadlines for the resolution of application review. The Program Comment, as drafted, would fail to provide the relief needed for the railroads to initiate the deployment of approximately 22,000 PTC-related wayside facilities to satisfy the Congressional public safety mandate, and will delay the installation of PTC past the Congressionally mandated 2015 deployment deadline.

In May 2013, at the FCC's request, the railroads suspended the submission of applications for Section 106 review for PTC wayside infrastructure while the Commission developed a "streamlined" solution for historic preservation review processing, effectively putting all PTC infrastructure deployment in the railroad rights of way on hold.[16] While the AAR appreciates the FCC's efforts to adopt a Program Comment to address PTC deployment, the historic preservation review process outlined in the draft Program Comment will take at least as long, if not longer, than the existing mechanism to review applications for PTC-related wayside deployment.

A. The Existing Section 106 Process Cannot Expedite the Processing of Large Deployments such as PTC

The current FCC historic preservation review process, which is set out in the Nationwide Programmatic Agreement ("NPA"),[17] suffers from several shortcomings that have an especially significant negative impact on PTC deployment:

- The Tower Construction Notification System ("TCNS") process is not suited for the review of relatively large numbers of applications submitted at the same time;
- There are no deadlines to provide a backstop for the resolution of Tribal review, and Tribal Nations may make automated information requests that have the effect of delaying approval;
- Consultative fees charged for Tribal review and monitoring are not transparent; and
- The process places an undue burden on consultative stakeholders to review applications for facilities with *de minimis* effects on historic properties, as evidenced by the lack of findings of significant impact or complaints with respect to historic reviews of wayside facilities completed to date.

The FCC has acknowledged that "[p]arties seeking to deploy wireless infrastructure often face processes they must complete prior to construction that can take long periods of time and impose significant expense."[18] The TCNS process was not designed for the approval of relatively large numbers of applications at the same time. The precipitating factor in the FCC's decision in May 2013 to halt the submission of PTC-related wayside infrastructure was the breakdown of the TCNS process following the submission of 280 applications by a railroad.[19] As the NPA provides that TCNS is the most definitive means for an applicant to ensure it has made contact with all relevant Tribal Nations, the failure of TCNS to accommodate the submission of several hundred applications at a time poses an obvious challenge for the processing of the 22,000 pending PTC wayside facilities.

More critically, the existing Section 106 process does not include deadlines to provide a backstop for the resolution of Tribal review, and Tribal Nations may make automated requests for additional information that can have the effect of postponing the resolution of approval for months after submission. In practice, although the NPA was implemented to improve the historic review process and streamline the approval of communications towers, the railroads have found that Section 106 review for an application typically takes at least five or six months from submission to approval, and can take as long as nine months.[20] While the NPA provides clear deadlines in cases where a Tribal Nation expressly disavows any interest in consultation, the FCC has taken the position that the Section 106 review process also allows a Tribal Nation to make automated requests for information to supplement the submission packet, and that each such request essentially "stops the clock" on Tribal review. The submission of an application packet often triggers an apparently automatic request for additional information from Tribal Nations served by TCNS, putting these applications in an administrative limbo with no clear path to resolution. In many cases more than one Tribal Nation expresses interest in consultation on a potential deployment site, giving rise to multiple opportunities for delay, but information requests on the part of only one Tribal Nation can significantly extend the duration of the approval process. As Verizon noted recently regarding this general problem with the current Section 106 process, "[i]f even one tribe does not respond to a notification or fails to render a determination about the effects of a project, the entire project will be delayed by a minimum of 60 days, but many times…the time is far longer."[21]

The current Section 106 process also involves highly variable and unforeseeable costs for applicants to gather data in response to Tribal requests, as well as to pay Tribal consultative and monitoring fees. Verizon reports that the cost for a consultant's report can be as much as $4700 per installation site, which does not include the cost of Tribal consultation fees and additional studies or tests.[22] For a proceeding of the scale of PTC, consultative fees add up quickly. After submitting fewer than 300 poles to the FCC for historic review processing, one railroad received requests from various Tribal Nations for the payment of consultation fees totaling $338,000, or $1,203 on average per site.[23] The lack of transparency regarding potential consultative fees and inability to predict the extent of information requests that will be made results in a difficult budgeting process for applicants for historic preservation review.

One of the most challenging aspects of the FCC's existing Section 106 process is it places serious burdens on Tribal Nations as consulting parties as well as on the applicants. As the NPA contains few exclusions, Tribes are required to review large numbers of applications with limited resources, even when the proposed facilities are likely to have little or no potential impact on any historic property of cultural and religious significance to any Tribal Nation. For example, the United South and Eastern Tribes ("USET") has described how Tribal Nations struggle under the annual burden of reviewing hundreds of communications tower applications, not to mention the added burden of having to review applications for 22,000 PTC-related wayside facilities.[24] The lack of exemptions in the current process for the deployment of infrastructure that will pose no or a *de minimis* risk of adverse effects on historic properties increases the burden on all applicants and stakeholders, including Tribal Nations, in the PTC- related infrastructure review process.[25] Based on the railroads' previous experiences with Section 106 review, however, the vast majority of communications infrastructure similar in size and location to the PTC-related wayside facilities are ultimately found to have no or minimal effect on historic properties, and so require no further assessment prior to implementation. The FCC should draw from its lengthy history of overseeing the clearance of structures similar to PTC-related facilities without the need for mitigation measures, and draft the Program Comment to avoid increasing the amount of historic preservation review that will ultimately end with a finding of no effects on historic properties.

B. The Process Outlined in the Draft Program Comment Will Not Provide the Needed Relief

The draft Program Comment was intended to address the deficiencies in the current historic review process regarding the expeditious review of PTC-related infrastructure, and the proposed SHPO exemptions will improve and expedite this side of the review process. The process outlined in the draft Program Comment, however, suffers from the following specific problems:

- The revised process will not apply to any PTC-related infrastructure deployed by at least three of the seven Class I railroads;
- While the reformed TCNS submission process purports to offer administrative convenience, it continues to anticipate time-consuming, pole-by-pole Tribal review as opposed to clearance of larger sections of track;

- Like the existing historic preservation review process, the draft Program Comment does not offer definitive deadlines for the resolution of Tribal review;
- Tribal Nations may continue to make automated information requests which have the effect of delaying approval;
- The addition of new required documents, including cultural resource reports, will increase the time needed to prepare applications for submission as well as the time needed for Tribal Nations to review applications, and producing such information is an inefficient and costly exercise as not all Tribal Nations have asked for such documentation; and
- The proposed process allows Tribal monitoring and/or alternative excavation techniques upon request for every pole deployment, without requiring any evidentiary showing of the probability of the presence of cultural resources, instead of as mitigation techniques where necessitated.

The draft Program Comment will not apply to all railroads' PTC-related wayside infrastructure. Although the railroads have previously shared their deployment plans with the FCC, the dimensions of the foundations that are set as the maximum limit for PTC-related facilities in the draft Program Comment are exceeded by at least three railroads.[26] The FCC and industry alike are not well served by a Program Comment that fails to encompass such a large number of PTC-related wayside facilities in its provisions.

Without batched review of applications, the draft Program Comment is of limited utility. In the draft Program Comment, the FCC provides that it may, at its discretion, permit the batched submission of applications for review of PTC-related infrastructure into TCNS. However, the Commission clarifies that any batching of submissions "is for administrative convenience and is not intended to affect the level of review."[27] Under the ACHP's rules, a Program Comment is intended to be used "[w]hen effects on historic properties are similar and repetitive or are multi- State or regional in scope," and Program Comments previously approved by the ACHP generally preclude case-by-case review of undertakings.[28] While the convenience of batched ***submission*** may be minimally useful in reducing the work of submitting thousands of separate PTC-related applications into TCNS, the failure of the Program Comment also to provide for batched ***review*** of PTC-related infrastructure effectively dooms the utility of the proposed process.[29] Other than the mild and, as the FCC acknowledges, purely administrative convenience of being able to enter groups of PTC-related wayside facilities applications, rather than having to enter each application separately, the batching provision offers no advantages to the railroads, and will not have any impact on the way Tribes will conduct their review.

The draft Program Comment does not include definitive deadlines for the resolution of Tribal review. Although the draft Program Comment facially offers some helpful shortened deadlines (sixty days to forty days) for Tribal Nations to determine that they are not interested in further consultation regarding a specific application, this small benefit is greatly outweighed by the fact that— like the existing Section 106 process—it does not include definitive deadlines for the resolution of Tribal review. Instead, Tribal Nations are permitted to follow up receipt of a submission package with requests for additional information, with no limit on the time to review information packages once provided. The ability to respond to an application with additional information could render the shortened forty day period meaningless, as review in fact could last indefinitely.

The draft Program Comment's deadline for dispute resolution of sixty days, "unless the FCC determines additional time is necessary," is similarly ineffective. Indeed, in the railroads' experience, even with significant additional time, the FCC has been unable to resolve a much smaller number of disputes involving TCNS entries of non-PTC rail infrastructure. The FCC is not adequately staffed now to resolve disputes between various stakeholders, and there is no reason to believe that it could satisfy a sixty day deadline when faced with the review of thousands of entries.

Although the draft Program Comment takes the unprecedented step of requiring the default submission of a cultural resource report for each PTC-related location prior to any request by a Tribal Nation, the utility of this provision is questionable if Tribal Nations have no deadlines for the review of such information. In the railroads' experience, preparation of the full list of required information documents, including cultural resource reports, will take from four to six weeks *per installation.* If adopted as drafted, the Program Comment could eliminate the possibility of the deployment of any PTC-related infrastructure in 2014 for some railroads, as clearances could come too late in the construction season to arrange for work crews before the winter season begins. No benefit will be gained if applicants are required to expend considerable resources to provide such studies upfront if such submissions are not required by a particular Tribal Nation or are not reviewed expeditiously.

Moreover, such a provision effectively inverts the Section 106 process by placing the burden of identifying historic properties of cultural and religious significance to Tribal Nations on the railroads.[30]

The draft Program Comment does not require an evidentiary showing prior to monitoring and/or alternative excavation. The draft Program Comment establishes the ability of Tribal Nations to request monitoring and/or alternative excavation techniques on request, with no evidentiary showing. This is a backwards approach to the Section 106 review process. Under the NPA, Tribal consultation is a two-part process. First, consultation is intended to ascertain whether any historic property of cultural and religious significance might be located within the Area of Potential Effects ("APE"). Second, if a determination is made that any such a historic property exists, consultation should attempt to reach an agreement on the presence or absence of effects on that property.[31] Only at this second stage would mitigation such as monitoring or alternative excavation techniques be necessary. In contrast, the draft Program Comment suggests that before a Tribal Nation has even ascertained that any historic property exists that could be subject to direct or visual effects from the PTC-related deployment, that Tribe can require that *all* sites be subject to monitoring or to alternative excavation. In fact, with no evidentiary showing, under the draft Program Comment any potentially interested Tribal Nation can require monitoring for any site even if a railroad has voluntarily commissioned a field study by a Secretary of the Interior-qualified professional archeologist who has made a determination that no archeological historic properties exist within the APE.[32] The draft Program Comment also does not address the logistics of coordinating the potentially dozens of Tribes who could claim an interest in sending a monitor to observe the installation of a particular site, including the challenge of organizing and scheduling deployments and ensuring the safety of all track-side personnel.

As discussed in Section IV below, to request monitoring or alternative excavation methods as a form of mitigation, a Tribal Nation should first be required to establish—based on its own records, historical documents, or specific cultural resources—that a historic property of cultural and religious significance exists within the relevant APE. As currently drafted the Program Comment establishes the entire national railroad rights of way as a historic property, subject to blanket mitigation. The railroad rights of way, which have been operated for up to 175 years, have been subject to significant, heavy construction and maintenance associated with railroad operations. At a maximum height of seventy-five feet, the PTC-related wayside facilities would be considerably smaller than standard communications towers, and would be closer in height to standard utility poles that are already ubiquitous in the urban and rural landscape. Due to the high level of previous disturbance on the railroad rights of way, and the small scale of the PTC-related wayside facilities, the likelihood of existing cultural resources that could be affected by PTC-related wayside infrastructure deployment and require any form of mitigation is minimal.[33] The railroads continue to be open to mitigation in any situation where a Tribal Nation expresses a specific concern regarding a historic property and provides evidence to support this concern.

III. The Program Comment Process Does Not Require That All Impacts on Historic Properties be Avoided

The draft Program Comment does not need to guard against any potential impact to any unknown historic property. Congress did not intend that the Section 106 process be exhaustive. As the Commission has previously noted, the NHPA "contemplates a balancing of the likelihood of significant harm against the burden of reviewing individual undertakings" and "does not require perfection in evaluating the potential effects of an undertaking in every instance."[34] Specifically, Section 106 and the ACHP's rules require that federal agencies "take into account" the effect of their undertakings on historic properties, and engage in a "good faith and reasonable effort" to identify historic properties.[35] The standard of review for undertakings under the Section 106 process "is not one of perfection but one of reasonableness, taking into account both the likelihood that adverse effects will not be considered in some instances and the overall benefits to be obtained from streamlining measures."[36]

One of the main reasons that the historic preservation review process outlined in the NPA provides for few exclusions is that the type of communications infrastructure that its drafters envisioned was considerably more intrusive on the human environment than PTC-related facilities. The FCC's environmental and historic preservation review rules were developed at a time when wireless infrastructure deployment generally meant the construction of a single 300 foot communications tower that loomed over a previously undeveloped greenfield.[37] In contrast, the wayside facilities that will be deployed by the railroads to enable PTC deployment will be considerably smaller, located in areas that have been subject to extensive soil disturbance and are used continuously for rail transportation purposes, and pose *de minimis* risk of negative effects on the human environment.[38]

As previously discussed, the railroads have seen firsthand that the overwhelming majority of wayside facilities similar in size to PTC-related deployments ultimately are determined by consulting parties to have no effect on any historic property, or to have such a *de minimis* effect that no mitigation is necessary. In fact, similarly-sized facilities on the railroad rights of way that have been processed by the railroads to date have been cleared without a finding of adverse impact. The FCC should give heavy weight to the lack of historic preservation concerns raised regarding all PTC-related facilities located on the railroad rights of way to date. Ultimately, the railroads are caught between two statutory mandates—the need to conduct historic preservation review and the need to satisfy the Congressional mandate for nationwide PTC deployment to meet significant national safety objectives. The AAR asks the FCC to use its considerable discretion in drafting the Program Comment to ensure that these statutory imperatives can be reconciled, so that PTC deployment may move forward. The current draft Program Comment will not accomplish this goal.

IV. The Program Comment Should Encompass All PTC-Related Facilities, Impose Firm Deadlines, Emphasize Tribal Responsibility to Identify Historic Properties, and Provide for Limited Monitoring

The AAR continues to feel strongly that the best solution to address the need for expedited Section 106 review of PTC-related infrastructure remains an exemption from SHPO and Tribal review for all wayside facilities no taller than seventy-five feet located on the railroad rights of way that are not immediately adjacent to any previously recorded historic properties.[39] Because of their small size, minimal area of direct and indirect impact, and location on previously disturbed industrial rail corridors, the potential effects of PTC-related facilities on historic properties are foreseeable and minimal or not adverse. The FCC has broad authority pursuant to the NHPA and the ACHP's regulations to seek a program alternative that would exempt most PTC-related wayside facilities from Section 106 review.[40]

If the FCC declines to adopt a general exemption for all PTC-related wayside infrastructure, it should draft the Program Comment to provide a process that is superior to the existing Section 106 application submission and review procedures. As an initial matter, the Program Comment should be drafted to provide relief to all railroads, and to exclude all PTC- related facilities located on the railroad rights of way from SHPO review. To initiate Tribal consultation, the only information (other than that required by the current TCNS process) that the railroads should have to submit is detailed maps that will allow Tribal consultants to determine whether the PTC-related facilities would have the potential to affect known historic properties of cultural and religious significance. These maps could be submitted through TCNS, or provided directly to interested Tribal Nations. Once the consultative process is initiated, the Program Comment should provide strict, binding deadlines to ensure that the application review will be resolved in a finite period. The failure to meet such deadlines should be construed as a lack of interest in further consultation. Requests for monitoring by Tribal Nations should only be granted on a showing of evidence that a historic property could be affected by the proposed deployment, with a limit of one monitor per work crew. More details of the AAR's proposed revisions to the draft Program Comment are provided in the attached Appendix.

Failing to make the recommended revisions to the draft Program Comment included below will almost certainly risk the loss of most, if not all, of another construction season this year and, in turn, the inability of the railroads to meet the end of 2015 deadline for nationwide PTC deployment mandated by Congress. According to FCC staff, the Commission plans to submit the draft Program Comment to the ACHP in March 2014, which should result in final approval of the Program Comment in late April or early May 2014.[41] Based on this timeline, and given the length of time needed for the preparation of cultural resource reports and other materials mandated by the current draft Program Comment, the railroads would not be able to submit their first applications for PTC-related wayside deployments until late June 2014. Even under the best circumstances as provided under the draft Program Comment, no PTC-facility would be likely to be approved for construction until forty days after submission—or mid-August 2014. Any request for additional information by a Tribal Nation would have the effect of significantly pushing back even these best-case scenario deployment dates, and could result in at least some railroads being unable to deploy any PTC-related facilities in 2014.

Although the Tribal Nation consultative process is not the only factor the railroads must take into account when planning for PTC deployment, delays in approval have already significantly compromised, and will continue to negatively impact, the rest of the construction and implementation process, including ordering and receiving PTC equipment and arranging for contractors for deployment services. For tracks on the northern plains in particular, the deployment season is normally limited by inclement weather by the early fall, and uncertainty regarding when approval for deployment could come will result in the inability to sign contracts to secure work crews before cold weather makes construction impossible. Being unable to deploy PTC wayside facilities in 2014 will also have a profound, negative impact on the testing of PTC systems, and will push back the training and certification of railroad employees on PTC equipment, which must take place before general deployment.

The Program Comment Should Apply to All PTC-Related Facilities. As drafted, the Program Comment is limited to infrastructure situated in a railroad rights of way supporting either a wayside antenna or base station that is no taller than seventy-five feet (including antenna), requires a foundation no deeper than fifteen feet, and creates a foundation hole not in excess of fifteen inches in diameter.[42] PTC-related infrastructure that falls outside of these categories must rely on the Section 106 review process established under existing FCC regulations and procedures. As has been previously disclosed to the FCC, at least three of the seven Class I railroads plan to deploy PTC-related facilities using a foundation hole that will exceed the fifteen inch diameter provided in the draft Program Comment. Rather than make the Program Comment process unavailable for almost half of the affected railroads, the AAR believes that the FCC can minimize ground disturbance by refining its definition of covered facilities.

While the diameter of poles deployed by most railroads will be approximately fifteen inches, at least one of the railroads will be deploying PTC wayside facilities using an augered foundation that is eighteen inches in diameter, with a disturbance diameter of up to twenty inches, while two other railroads will be using foundations that will be wider than those contemplated in the Program Comment but also very shallow, requiring a foundation that is less than six feet deep. The FCC can revise its constraints regarding foundation dimensions to include the deployment plans of all railroads without giving rise to any increased risk of adverse effects on historic properties. To ensure that no eligible PTC-related infrastructure is excluded from the Program Comment process, the FCC should clarify that foundation deployments should either have a disturbance diameter of no more than twenty inches with a foundation depth of no more than fifteen feet, *or* an open excavation of any size with a foundation less than six feet deep. Such a provision will allow all of the railroads to rely on the Program Comment while not increasing any potential impact to the human environment.

Tribal Nations Should Be Encouraged to Exclude Counties from Section 106 Review. As drafted, the Program Comment would exclude from SHPO review facilities constructed in existing railroad rights of way where similar structures already exist in the same vicinity.[43] The AAR appreciates this helpful exclusion. However, the Program Comment would provide no parallel exclusion from review by Tribal Nations, despite the fact that Tribes and SHPOs share similar historic preservation concerns.[44] The AAR believes that Tribal Nations should be able to designate areas, such as counties, for which they are not interested in consultation, and to provide those designations to the railroads as an effective exemption from review.[45] To clarify this provision, the FCC should draft the final Program Comment to exclude from Tribal Nation review all PTC-related wayside facilities that have been designated as including no historic properties of cultural or religious significance.

Applicants Should Not Be Required to File Cultural Resource Reports with TheirApplication. The draft Program Comment provides that applicants seeking to use the FCC's revised historic review process must submit a cultural resources report, prepared by a professional who meets the relevant standards in *The Secretary of the Interior's Professional Qualifications Standards*, with their application.[46] The FCC acknowledges that ordinarily applicants are not required to provide such reports as part of their TCNS submissions, but suggests that requiring such a submission is necessary to expedite the Tribal review process, as Tribal Nations "often request a cultural resources report" on receiving an application via TCNS. The AAR believes that rather than asking the railroads to spend the extensive time needed to prepare a cultural resource report for each PTC-related pole or facility, when Tribal Nations are already facing the challenge of reviewing thousands of pole applications, the resources of all stakeholders would be better spent preparing and reviewing the detailed maps which are also required. Such maps provide all information needed to assess whether a particular deployment will have a potential effect on a known historic property of cultural and religious significance to that Tribal Nation.

The requirement of Tribal consultation under Section 106 is based on the presumption that Tribal Nations are better suited to identify historic properties of cultural and religious significance to that Tribe than are applicants or Federal agencies. For this reason, a Federal agency's duty under the NHPA and the ACHP's rules is to seek and consider information regarding historic properties from Tribal Nations. As the NPA notes, the purpose of preliminary communications between an applicant and Tribal Nation is "to ascertain whether Historic Properties of religious and cultural significance to the Indian Tribe…may be affected by the undertaking and consultation is necessary."[47] There is no corresponding requirement that the agency, or an applicant, convey all known or suspected information to the Tribes.[48] Moreover, many of the required contents of the cultural resources report as provided in the draft Program Comment do not relate to Tribal interests at all, such as "[i]nformation on Federal lands…along or under tracks" and "[o]wnership of tracks on or near Federal lands, including direct ownership or lease arrangements," and so will be of limited utility in assessing whether a historic property of cultural and religious significance to a Tribe might be affected.[49] For the railroads, the resources needed to assemble cultural resources reports for each PTC-related facility will be overwhelming. As discussed above, on average, the preparation of such a report takes anywhere from one month to six weeks prior to submission per installation.[50]

The railroads are also concerned that although the submission of cultural resource reports for each PTC-related facility is intended to expedite review, the draft Program Comment provides a mechanism for Tribal Nations to respond to the receipt of such reports with additional information requests. At least one railroad that has taken advantage of recent permission to resume limited submission of PTC-related infrastructure in TCNS has had a cautionary experience.[51] Despite submitting a supplementary information package that contained many of the documents described in the draft Program Comment, the railroad continued to receive automated responses from Tribal Nations requesting the submission of an information package. As the railroad had, in fact, submitted such a package, it was impossible to know if the package had been reviewed, what additional information these Tribal Nations might need, or how the railroad was to make a determination regarding how to supplement its submission. If Tribal Nations can respond to a cultural resource report with requests for additional information that can lead to infinite delays in approval, there are no efficiencies gained by generating such reports in advance of Tribal requests. The FCC should draft the Program Comment to eliminate the ability of Tribal Nations to respond to an application with an automated request for additional information.

Deadlines. One of the most fundamental problems with the existing Section 106 review process is the lack of clear and finite deadlines for approval. The draft Program Comment shortens the potential approval period from approximately sixty days to forty days, but does not the fundamental problem of unlimited consultative review by Tribal Nations.

As drafted, the Program Comment provides that if an applicant has not received a response from a Tribal Nation twenty days after the application was submitted via TCNS, provided the applicant has attempted at least one follow-up communication during that period, the railroad may ask the FCC to send a letter to the Tribal representative.[52] The FCC will send this letter within five days of the request, and if the Tribal Nation does not respond within fifteen days, it will be deemed to have no interest in consultation.[53] The AAR believes that an approximately forty day period for the approval of PTC-related infrastructure is appropriate, and supports this general timeline provision.

However, there are many circumstances under which the Tribal review process is essentially open-ended. Notably, during the initial twenty day period, a Tribal Nation may ask for additional information. If the railroad provides such information, the draft Program Comment provides no deadline during which the Tribal Nation must complete the review. If the Tribal Nation and the railroad disagree about any aspect of this request, after attempting to resolve their differences in fifteen days the parties may bring their disputes before the FCC. The draft Program Comment does not provide any timeline for the ultimate resolution of such a dispute by the FCC. Similarly, although the FCC pledges to resolve all disputes regarding a submission that requires further review, or a closer examination, within sixty days, it caveats this provision by noting that it can take additional time to resolve the dispute if it determines this is necessary.

Given the narrow questions and largely uniform nature of the PTC-related wayside facilities for review, the AAR believes that all disputes regarding any PTC-related pole should be definitively resolved within thirty days. Based on the detailed maps supplied by the railroads, a Tribal Nation should be able to make a determination regarding the likelihood of historic properties of cultural and religious significance at each pole location within that time period. Moreover, the FCC has ample authority to impose firm deadlines on Tribal Nations for the completion of Section 106 review. The ACHP's rules provide that an agency need only provide a Tribal Nation a "reasonable opportunity" to identify concerns regarding historic properties,[54] and the 2000 Executive Order establishing guidelines for consultation and coordination with Tribal Nations provides that agencies must only establish procedures that allow for "timely input" by Tribal officials.[55] Other agencies have imposed firm deadlines on Tribal Nations,[56] and in rare challenges courts have affirmed that "agencies…[may] set deadlines as needed in order to ensure the timely and proper disposition of matters" before them.[57]

If a Tribal Nation makes a determination, accompanied by an evidentiary showing, that a historic property could be affected by a particular PTC-related facility, it should be able to request the presence of a monitor during installation or an alternative excavation method (*see discussion below*). If the Tribal Nation finds that there is no such historic property of cultural and religious significance, or that it will not be affected by the undertaking, the deployment of PTC-related infrastructure should be allowed to proceed. The failure of a Tribal Nation to respond to an application, or to meet any consultative deadline, should be construed as an expression by that Tribe that it has no interest in review of the proposed facility.

Appropriate Limits Should Be Established for Tribal Monitoring and Requests for Alternative Excavation. The draft Program Comment provides that "a Tribal Nation may request to monitor construction at *any or all sites* within a batched submission."[58] The AAR believes that providing pole-by-pole monitoring as a default will entirely defeat the purpose of the Program Comment, which is to provide broad, systematic relief for the deployment of all PTC- related wayside facilities. Allowing individual monitoring of all PTC-related poles represents no improvement over the current Section 106 review process. Monitoring, as a form of mitigation, should only be invoked if the Tribe "provides evidence that supports a high probability of the presence of intact archeological Historic Properties within the APE for direct effects."[59] Any disputes arising from the submission of such evidence, including the sufficiency of such evidence, and any disputes regarding whether requests for mitigation or alternative excavation methods should be honored, should be resolved by the FCC within fifteen business days.

Where monitors have been shown to be appropriate, for safety reasons the Program Comment should clarify that a maximum of one monitor will be allowed per railroad work crew.[60] The AAR believes that the most comprehensive approach to monitoring would be the formation of a pool of professionals who satisfy the Secretary of the Interior's Professional Qualification Standards. Both the railroads and Tribal Nations should be able to contribute monitors to this pool. The railroads would then draw from this group of approved monitors to accompany work crews installing any PTC-related wayside infrastructure for which a Tribal Nation had provided evidence of potential archeological impact. Any disputes regarding the selection of a monitor for areas of interest for more than one Tribal Nation should be submitted to the FCC, and resolved within fifteen days. The largely uniform historic preservation interests of all Tribal Nations will be well-represented by a Secretary-qualified monitor, and providing for a single monitor will avoid inevitable scheduling delays and safety concerns that will arise if each interested Tribe is entitled to deploy a monitor to each PTC-related wayside site.

As a related concern, the AAR disagrees with the provision in the draft Program Comment that provides that a railroad must honor any request by a Tribal Nation to use an excavation method other than screwing in of the pole or auger drilling "[w]here necessary to ascertain the presence of archeological resources."[61] The railroads have invested significant resources in PTC deployment, which includes making a determination regarding the type of foundation and make-up of crews needed to effect this major public safety infrastructure project. Because of the significant financial cost and strain on resources, and safety risks, a request for monitoring or alternative excavation should only be acted upon with a showing by the Tribal Nation that a known historic property of cultural and religious significance exists within the APE, and that the PTC-related wayside facility could negatively affect such a property absent mitigation.

V. The Section 106 Review of Previously Constructed PTC Facilities Should be Resolved Separately

The final section of the draft Program Comment addresses previously constructed PTC facilities, and provides that in order to "benefit from the efficiencies" of the revised historic review procedures, a railroad must provide complete responses to all information requests from the FCC regarding previously constructed PTC facilities. The railroads fully intend to comply, in a timely manner, with all FCC requests. Since Congress first imposed the PTC mandate, the railroads have been trying to implement an important public safety mandate as quickly as possible, with no intent to circumvent existing rules or processes.[62] The railroads have always believed that there would be no significant environmental impact or impact on any historic properties from the deployment of infrastructure on its rights of way. While the railroads are prepared to cooperate with the FCC and the Tribal Nations to address concerns regarding previously constructed PTC facilities, to the extent the draft Program Comment suggests otherwise, resolution of those matters cannot be a condition for a workable Section 106 process that permits the timely installation of the thousands of remaining wayside structures required for PTC.

VI. Conclusion

Rather than representing a streamlined solution to the existing historical review process, the proposed Program Comment would perpetuate the extensive delays that are characteristic of that process. In particular, the lack of definitive deadlines for the resolution of Tribal consultation only perpetuates problems already inherent in the FCC's Section 106 process. The AAR continues to believe that given the low probability of significant environmental or historic impact, as evidenced by the experience of the railroads with the Section 106 process to date, the FCC should seek an exemption from Section 106 review for all PTC wayside facilities no more than seventy-five feet in height that are located on the railroad rights of way and not within or immediately adjacent to a known, previously recorded historic property. Absent an exemption for all PTC deployment, the FCC should revise the draft Program Comment to encompass all PTC-related facilities, impose clear deadlines for the resolution of applications, emphasize Tribal responsibility to identify historic properties, and provide for monitoring and alternative excavation methods only as a form of mitigation when the potential for adverse effects has been established.

Respectfully submitted,
By: */s/ Michele C. Farquhar*

Louis P. Warchot
Senior Vice President-Law and General Counsel
Timothy J. Strafford
Assistant General Counsel

The Association of American Railroads
425 3rd Street, S.W. Suite 1000
Washington, DC 20024 (202) 639-2502
Michele C. Farquhar Deborah K. Broderson Hogan Lovells US LLP

Counsel to the Association of American Railroads
555 13th Street, NW Washington, DC 2004 (202) 637-5663

February 12, 2014

APPENDIX. SUMMARY OF THE AAR'S PROPOSED PROGRAM COMMENT PROCESS FOR PTC

The AAR continues to believe that the best solution to address the need for expedited, comprehensive Section 106 review of PTC deployment is an exemption for all such facilities no more than seventy-five feet in height that are located on the railroad rights of way and not within or immediately adjacent to a known, previously recorded historic property. If the FCC believes that a Program Comment is necessary, the AAR recommends that the Commission adopt the historic review process outlined below.

- General
  - The Program Comment creates an optional alternative process to the existing FCC Nationwide Programmatic Agreement (NPA)
  - Rather than relying on the Program Comment process, in the alternative railroads may use the existing Section 106 process outlined in the NPA, or may enter into arrangements or agreements with Tribal Nations governing the review of all PTC facilities
- Applicability
  - Includes all PTC-related facilities located in the railroad rights of way
  - Such facilities must not:
    - Be taller than 75 feet (including antenna);
    - Result in a foundation hole that has a disturbance diameter of more than 20 inches with a foundation depth of more than 15 feet, *or* an open excavation of any size with a foundation more than 6 feet deep; or
    - Be situated outside the railroad rights of way
  - For all other cases, including collocations, Section 106 review will be conducted under existing FCC regulations and procedures

- The Program Comment does not govern any Section 106 responsibilities agencies other than the Commission may have with respect to those agencies' federal undertakings, but the Program Comment may be adopted by other Federal agencies to satisfy their obligations under Section 106
- The Program Comment does not apply on Tribal lands unless a Tribal Nation elects to adopt its provisions and in so doing follows the process set forth in Section I.D of the FCC NPA
- Exclusions from Section 106 review
  - All PTC-related facilities (as defined above) that are similar to nearby existing structures within the existing railroad rights of way, provided the location is not within the boundaries of a known historic property, and including those areas designated by Tribal Nations as not of consultative interest; and
  - Effects on the rails and the track beds themselves
- Applicant submission package
  - May either use TCNS and E106 (where applicable)
  - May batch the submission of poles by county
    - Batched submissions will be accompanied by a detailed map providing locations
  - The submission of a cultural resources report, field survey and/or ethnographic survey is discretionary
- Area of Potential Effects (APE)
  - Presumed ¼-mile APE for visual effects
  - Linear APE recommended
- Tribal Nations Review
  - Scope of review: Tribal Nations will review the maps submitted by the railroads to determine whether a facility site is located within a known area of cultural and religious interest
  - Compensation: Tribal Nations may request compensation for the review of applications consistent with the scope described above
    - For areas determined by the FCC to have a high probability of the presence of intact archeological Historic Properties within the APE for direct effects and for which a Tribal Nation has requested a monitor, the Tribal Nation may request compensation for the monitor provided that the monitor meets the criteria described below
  - Tribal Nations may request compensation for the review of applications consistent with ACHP guidance, including recovery

of expenses to retain additional human resources to review submitted maps for known areas of cultural and religious interest

- Applicants shall ask Tribal Nations whether any clarification regarding the provided maps is needed within 20 days of submission
    - The ability to use the automated response in TCNS to request additional information will be disabled for the Program Comment process
- If a Tribal Nation does not respond to the TCNS submission within 20 days, within which applicant has attempted at least one follow-up contact:
    - The applicant may ask the FCC to send a letter and/or e-mail to the Tribal Nation's designated cultural resources representative seeking a response
    - The FCC will send this letter or e-mail within 5 business days of the applicant's request
    - If the Tribal Nation does not respond within 15 business days after the FCC has mailed its letter or e-mail, it will be deemed to have no interest in review of the proposed facility
- For those Tribal Nations that have notified the FCC that they may generally be considered uninterested in TCNS submissions if they do not respond within a specific time period of 30 days or less, without any need for follow-up contact, the usual process applicable to those Tribal Nations shall apply
- If a Tribal Nation feels that the information provided by the applicant (*e.g.* Google Earth overlays on U.S. Geological Survey maps) is insufficient, the Tribal Nation may appeal directly to the FCC, which will resolve any disputes within 15 business days
- The FCC will resolve all other disputes between the applicants and Tribal Nations (other than disputes regarding the monitoring process; see below) within 30 business days of a request from either or both parties to intervene
- The FCC has full discretion to intervene in Section 106 review at any point in the process

- SHPO Review
    - Applicant shall ask SHPO whether additional information is needed no later than seven days after submission
    - SHPO review should be completed within 30 days of submission pursuant to the procedures in the FCC NPA

- Addressing Adverse Effects
    - Processes in the FCC NPA for avoidance, minimization and mitigation continue to apply
    - If a Tribal Nation fails to meet any consultative deadline, as discussed above, it will be deemed to have no interest in reviewing the proposed facility
    - A Tribal Nation may request to monitor construction at any site for which it provides to the FCC evidence that supports a high probability of the presence of intact archeological Historic Properties within the APE for direct effects
        - The sufficiency of such evidence will be determined by the FCC within 15 business days of its submission by the Tribal Nation
    - Only one monitor will be staffed per track crew, and the monitors will be taken from a pool, created by the applicants and Tribal Nations, of professionals who meet the Secretary of the Interior's Professional Qualification Standards
        - Tribal Nations may be compensated for Tribal monitors who meet the Secretary of the Interior's Professional Qualification Standards and that are used for such mitigation
        - Any disputes relating to the selection of monitors shall be referred to the FCC, which will make a decision within 15 days

End Notes

[1] Originations do not exactly equal terminations because some crude oil that originates on U.S. Class I railroads might be delivered to U.S. short lines or to railroads in Canada for termination and because some crude oil that terminates on U.S. Class I railroads might originate on railroads in Canada or on U.S. short line railroads.

[2] A train-mile is the movement of a train the distance of one-mile. For perspective, a million train-miles is roughly equivalent to 300 train trips across the country.

[3] In terms of rail cars, "truck" refers to the complete four-wheel assembly that supports the car body.

[4] In the United States, FRA regulations specify that freight rail trackage be classified into one of six track "classes" based on track quality. The class of a section of track determines the maximum FRA-allowable speed for that section. Freight train speed on "excepted" and "Class 1" track cannot exceed 10 mph; on Class 2 track, 25 mph; on Class 3 track, 40 mph; on Class 4 track, 60 mph; and on Class 5 track, 80 mph. In practice, it is common for railroads, for a variety of reasons, to operate at speeds lower than the FRA maximum. (For example, U.S. freight trains rarely, if ever, exceed 70 mph, even on the best Class 5 track.) If a track's class does not permit 40 or 50 mph crude oil trains, those trains will travel no

faster than the FRA limit. Our understanding is that, on U.S. Class I railroads, most of the trackage on which trains with large amounts of crude oil travel are Class 4 or 5 tracks.

[5] Track geometry includes such parameters as track gauge, curvature, alignment, profile, and the cross level of the two rails. Track geometry inspections are generally performed by sophisticated stand-alone cars that use a variety of sensors, measuring systems, and data management systems to create a representation of the track being inspected.

[6] There may be cases where safety considerations (e.g., a track located on a grade) might dictate otherwise. In these cases, detectors will be installed as close to 40 miles as conditions allow.

[7] Some railroads also support 9/16th inch tank car shells for freight cars carrying ethanol.

[8] "An Overview of Unconventional Oil and Natural Gas: Resources and Federal Actions," Congressional Research Service, January 23, 2014.

[9] Some of you may have seen an article in the January 29 *Washington Post* on delays to the Washington metro's new "Silver Line." The article points out that the line's automatic train control system has delayed the completion of the project for months and still is not working correctly. The metro's train control system is many orders of magnitude less complex than what freight railroads are implementing. Still, the metro's problems offer a clear example of why the rail industry — and policymakers — should be very concerned with PTC implementation and the importance of making sure that such a complex system operates as intended, no matter how long it takes. Unlike the Washington metro, railroads do not have the luxury of simply not operating a new line.

[10] The Association of American Railroads ("AAR") is a voluntary non-profit membership organization whose freight railroad members operate 82 percent of the line-haul mileage, employ 95 percent of the workers, and account for 97 percent of the freight revenues of all railroads in the United States. More information on the AAR is available at our website, https://www.aar.org/Pages/Home.aspx.

[11] *Comment Sought on Draft Program Comment to Govern Review of Positive Train Control Facilities under Section 106 of the National Historic Preservation Act*, WT Docket No. 13-240, Public Notice, DA 14-97 (rel. Jan. 29, 2014) ("*Public Notice*").

[12] *See Public Notice* Attachment A ("*Draft Program Comment*").

[13] *See Public Notice* at 1; 16 U.S.C. § 470v.

[14] *See* Comments of the Association of American Railroads, *Comment Sought on Scoping Document for Development of a Proposed Program Comment to Govern Review of Positive Train Control Facilities under Section 106 of the National Historic Preservation Act*, WT Docket No. 13-240 (filed Nov. 15, 2013) ("AAR Scoping Document Comments").

[15] The AAR's recommendations for modifying the draft Program Comment are set out in the attached Appendix.

[16] With the permission of the FCC, since May 2013 some railroads have submitted some requests for approval using the established Tower Construction Notification System ("TCNS") process for the review of small base station towers that would otherwise qualify as undertakings covered by the draft Program Comment. On January 8, 2014, the FCC reopened TCNS on a limited basis in a Beta test format for the filing of PTC-related wayside facilities applications batched in groups of no more than twenty contiguous sites located within a single county in areas that were subject to discussion with Tribal Nations at meetings in Rapid City, South Dakota and Tulsa, Oklahoma in fall 2013. *See* Letter from Jeffrey Steinberg, Deputy Chief, Spectrum and Competition Policy Division, Wireless Telecommunications Bureau, Federal Communications Commission to Timothy Strafford, Assistant General Counsel, Association of American Railroads at 1-2 (Jan. 8, 2014) ("Steinberg Letter"). Several of the railroads have submitted applications for review for PTC-related facilities in the weeks since, and other railroads are engaged in the process of preparing submissions under this interim program. There have been increasing indications, however, that numerous Tribal Nations will not review applications for PTC- related wayside facility installations under the interim program.

[17] *See* 47 C.F.R. Part 1, App. C—Nationwide Programmatic Agreement Regarding the Section 106 National Historic Preservation Act Review Process ("*NPA*").

[18] *Acceleration of Broadband Deployment by Improving Wireless Facilities Siting Policies; Acceleration of Broadband Deployment: Expanding the Reach and Reducing the Cost of Broadband Deployment by Improving Policies Regarding Public Rights of Way and Wireless Facilities Siting; Amendment of Parts 1 and 17 of the Commission's Rules Regarding Public Notice Procedures for Processing Antenna Structure Registration Applications for Certain Temporary Towers; 2012 Biennial Review of Telecommunications Regulations*, WT Docket No. 13-238, WC Docket No. 11-59, RM-11688 (terminated), WT Docket No. 13-32, Notice of Proposed Rulemaking, 28 FCC Rcd 14238, 14240 ¶ 3 (2013) ("*Wireless Tower Siting NPRM*").

[19] *See* Letter from Theodore K. Kalick, Senior U.S. Regulatory Counsel, Canadian National Railway, to Stephen G. DelSordo, Federal Preservation Officer, Federal Communications Commission at 6 (May 9, 2013) ("Kalick Letter"), attached to AAR Scoping Document Comments.

[20] Similarly, Verizon recently conducted a survey of its Distributed Antenna System ("DAS") projects, and found that the average time to complete a review was 84 days, with poles requiring approval from multiple Tribal Nations potentially taking much longer. For example, Verizon reports that a DAS installation on the roof-top of a building in Pennsylvania with no historic effects required consultations with nine Tribal Nations, and the last response was received 126 days after the Tribal review process was initiated; while the installation of a similar small antenna in Cleveland, Ohio was approved by the SHPO in 37 days, but took 150 days to receive approval from all Tribal Nations contacted through TCNS. *See* Comments of Verizon and Verizon Wireless, *Acceleration of Broadband Deployment by Improving Wireless Facilities Siting Policies et al.*, WT Docket No. 13-238, WC Docket No. 11-59, RM-11688, WT Docket No. 13-32 at 9 (filed Feb. 3, 2014) ("Verizon NPRM Comments").

[21] *See id.* at 20.

[22] *See id.* at 9.

[23] *See* Kalick Letter at 6. More recently, a railroad submitting applications using the Beta TCNS process was informed by a consulting Tribal Nation that its fees for clearing PTC-related infrastructure would be double the fees for reviewing non-PTC infrastructure.

[24] *See* Reply Comments of USET, *Nationwide Programmatic Agreement Regarding the Section 106 National Historic Preservation Act Review Process*, WT Docket No. 03-128 at 2 (filed Sept. 8, 2003).

[25] *See Save Our Heritage, Inc., et al., v. FAA*, 269 F.3d 49, 58 (1st Cir. 2001) (affirming that an agency can exclude undertakings from both environmental and historic preservation review based on a finding of a *de minimis* effect on the human environment); *see also See Wireless Tower Siting NPRM*, 28 FCC Rcd at 14243 ¶ 11; *see also* Comments of the Association of American Railroads, *Acceleration of Broadband Deployment by Improving Wireless Facilities Siting Policies et al.*, WT Docket No. 13-238 *et al.* at 13-17 (filed Feb. 3, 2014) ("*AAR Wireless Tower Siting Comments*") (discussing authority of the FCC to exclude wayside deployments from historic preservation review based on a finding of *de minimis* effect).

[26] *See Section IV infra.*

[27] *See Draft Program Comment* at 7.

[28] *See* 36 C.F.R. § 800.14(b)(1); *see also, e.g.*, Program Comment Issued for Streamlining Section 106 Review for Actions Affecting Post-1945 Concrete and Steel Bridges, 77 Fed. Reg. 68790, 68791 (2012) (providing for programmatic, rather than site-by-site, review of undertakings affecting highway bridges).

[29] Moreover, the FCC has failed to address continuing technical shortcomings with the TCNS process. According to conversations with FCC staff, the only change made to TCNS since its use for PTC-related facilities was suspended in May 2013 has been an expansion in the

number of characters that can be entered in the one available free text field (the address field). It is not clear that the design issues that caused the FCC to instruct the railroads not to use TCNS have been resolved. Further, in its January 8, 2014 limited reopening of TCNS on a test basis, the FCC requested that the railroads provide more robust information packages beyond that currently required and submit them into TCNS. *See* Steinberg Letter at 2. While the railroads have complied by accompanying their applications with a supplemental information package, they have continued to receive automated responses from Tribal Nation requesting more information. This has created confusion, as it is unclear whether or not the submissions have been reviewed, and raises concerns that a similar provision in the draft Program Comment will not be effective in shortening the period for Tribal review.

[30] *See NPA* at Section VI.D.1.b (providing that applicants "shall gather information *from* Indian tribes...to assist in identifying Historic Properties of religious and cultural significance to them") (emphasis added).

[31] *See NPA* at Section IV.G.

[32] *See id.* at Section VI.D.2.

[33] The NPA provides that to assess potential effects on historic properties, applicants should consider factors such as "topography, vegetation, known presence of Historic Properties, and existing land use." *See NPA* at Section VI.E. In the case of deployment of PTC-related facilities on railroad rights of way, the number of known Historic Properties will be minimal, all vegetation has long been cleared from the ballasted track bed, and the land has been in use for the industrial deployment of rail lines for decades, and in many cases for over a century.

[34] Nationwide Programmatic Agreement Regarding the Section 106 National Historic Review Process, *Report and Order*, 20 FCC Rcd 1073, 1087 ¶ 35 (2004) ("*NPA R&O*").

[35] *See* 16 U.S.C. § 470f; 36 C.F.R. § 800.4(b)(1).

[36] *NPA R&O*, 20 FCC Rcd at 1082 ¶ 21.

[37] *See Wireless Tower Siting NPRM*, 28 FCC Rcd at 14243 ¶ 11; *see also* AAR Wireless Tower Siting Comments at 6-8.

[38] *See* AAR Wireless Tower Siting Comments at 11 (discussing the location of undertakings along transportation corridors as a critical factor in the ACHP's prior approval of categorical exclusions from Section 106 review).

[39] *See* AAR Scoping Document Comments at 11-13.

[40] *See id.* at 18-24.

[41] The ACHP has forty-five days from the date of receipt of the draft Program Comment to act on that program alternative by either adopting it, declining to comment, seeking additional information, or asking for an extension. *See* 36 C.F.R. § 800.14(e)(5).

[42] *See Draft Program Comment* at 5.

[43] *See id.* at 6.

[44] *See id.*

[45] *See NPA* at Section VI.B (providing that a SHPO or Tribal Historic Preservation Officer "may specify geographic areas in which no review is required for direct effects on archeological resources or no review is required for visual effects").

[46] *See Draft Program Comment* at 8.

[47] *NPA* at Section IV.G; *see also NPA* at Section VI.D.1.b.

[48] *See Slockish v. U.S. Federal Highway Admin.*, 2012 WL 3637465 *9 (D.Or, June 19, 2013).

[49] *See Draft Program Comment* at Appendix.

[50] In addition, the FCC staff have expressed concerns that the resources of the Tribal Nations are already stretched thin by the need to review nearly 22,000 applications, and at recent consultative meetings in Rapid City, South Dakota and Tulsa, Oklahoma some Tribal representatives stated that they preferred not to receive such reports, and lacked the resources to review such submissions.

[51] *See generally* Steinberg Letter (describing the Beta TCNS submission process).

[52] *See Draft Program Comment* at 10.

[53] *See id.*

[54] 36 C.F.R. § 800.2(a)(4).

[55] Executive Order 13175: *Consultation and Coordination with Indian Tribal Governments*, 65 Fed. Reg. 67249 (2000). Courts have found that an agency may prescribe any reasonable perimeters for Tribal consultation, as long as it abides by those guidelines. *See, e.g., Lower Brule Sioux Tribe v. Deer*, 911 F. Supp. 395, 397 (D.S.D. 1995) (finding that an agency could have satisfied its obligation with even "perfunctory" consultation, as long as this was in accordance with that agency's policies).

[56] *See, e.g.*, 25 C.F.R. § 262.3(b)(1) (providing that a Tribal representative reply to a request for information in thirty days); 25 C.F.R. § 262.8(c) (allowing a government official to act if a Tribal government has not responded to a request in fifteen working days); 43 C.F.R. § 7.7(a) (requiring notice of "at least" thirty days to a Tribe prior to the issuance of a permit that "may result in harm to, or destruction of, any Indian tribal religious or cultural site on public lands").

[57] *Fallon Paiute-Shoshone Tribe v. U.S. Bureau of Land Management*, 455 F.Supp. 2d 1207, 1220 (D.Nev. 2006).

[58] *See Draft Program Comment* at 11 (emphasis added).

[59] *See NPA* at Section VI.D.2.d.

[60] The AAR notes that the draft Program Comment does not provide criteria to govern the coordination of monitors and work crews. Such criteria are critical to ensure that PTC deployment can go forward as scheduled even if, for example, an appointed monitor fails to appear at a work site on a scheduled deployment date. Standard criteria are also necessary to ensure the safety of work crews and monitors on the job site. If the ACHP adopts a Program Comment that includes provisions for monitoring, the FCC should also finalize a list of required working criteria that would ensure that monitoring does not slow the PTC deployment process, or endanger the safety of that deployment.

[61] *See Draft Program Comment* at 11.

[62] In the draft Program Comment, the FCC references Section 110(k) of the NHPA, which provides that an agency should not grant a license to an applicant who intentionally significantly adversely affected a historic property. *See id.* (*citing* 16 U.S.C. § 470h-2(k)); *see also NPA* Section X. The railroads object to any implication that their actions regarding the previously constructed facilities represented an "intentional" attempt to violate Section 106 or any of the FCC's environmental or historic preservation review rules.

In: Transporting Crude Oil by U.S. Rail
Editor: Rosario S. McLaughlin
ISBN: 978-1-63117-837-5

Chapter 7

TESTIMONY OF BOB GRECO, GROUP DIRECTOR OF DOWNSTREAM OPERATIONS, AMERICAN PETROLEUM INSTITUTE. HEARING ON "ENHANCING OUR RAIL SAFETY: CURRENT CHALLENGES FOR PASSENGER AND FREIGHT RAIL"*

Chairman Blumenthal, Ranking Member Blunt, and members of the subcommittee, thank you for the invitation to testify on the important and timely topic of enhancing rail safety.

My name is Bob Greco, Group Director of Downstream Operations at the American Petroleum Institute. API represents all segments of the oil and natural gas industry, with more than 590 members who supply most of the nation's energy.

The revolution in North American energy development has been made possible by technological breakthroughs in decades-old methods of energy development, which has set this nation on the path to energy security, a concept unthinkable just a few years ago.

The energy policies we choose today will determine if our nation will continue its march toward global energy leadership; a unique and once in a generation opportunity. America's dramatic increase in domestic energy

---

* This is an edited, reformatted and augmented version of testimony presented March 6, 2014 before the Senate Committee on Commerce, Science, and Transportation, Subcommittee on Surface Transportation and Merchant Marine Infrastructure, Safety, and Security.

production has fundamentally altered the global energy markets and more broadly the geopolitical landscape for decades to come, all while providing a much needed boost to our economy.

In order to achieve our nation's full potential as a global energy leader, all of us have to work together to ensure that our energy infrastructure is capable of safely, reliably, and efficiently transporting ever-increasing amounts of domestically produced energy, whether by truck, barge, pipeline, or the focus of today's hearing, rail.

Meaningful and lasting improvement in rail safety will only come from a holistic and collaborative approach to accident prevention, mitigation and response, and the oil and natural gas industry will continue to work with our colleagues in government, the rail industry, and others on continual safety improvement.

The Memorandum of Understanding recently released between the railroad industry and the Department of Transportation, which outlines operational changes to improve rail safety, is an important step in our shared goal of improving the safety of America's freight rail system.

While it's true that 99.997 percent of hazardous materials transported by rail reach their destinations without incident ... the oil and natural gas industry is committed to getting to zero rail incidents. Because when it comes to safety the only number that matters is zero incidents.

Getting to zero will take the long-term commitment to working collaboratively with all stakeholders and applying all of our best science, research, and real-world data in a thoughtful and deliberate manner.

Being a safe steward of our nation's energy resources and providing leadership in raising the bar on industry performance isn't new to the oil and natural gas sector. For 90 years, API has been the world leader in developing and improving equipment and operating standards – 600 and counting – for oil and natural gas through a collaborative process that involves all stakeholders, as well as government regulators. This process is accredited by the American National Standards Institute, or ANSI, which is the same organization that accredits the government's national laboratories.

We have already assembled the best experts from our industry, the railroads, scientists and engineers and others to tackle some of the issues raised by the recent rail incidents involving the transport of crude oil. PHMSA also committed to join our effort to develop a comprehensive standard that addresses the classification of crude oil to ensure we are moving that product in the safest manner possible. This includes possible safety improvements for material characterization, transport classification, and quantity measurement of crude oil based on the best available science and data.

This is part of our industry's longstanding commitment to safety. In 2011, the oil and natural gas industry helped lead the multi-industry effort that led to significant improvements in the design of tank cars. And we moved forward voluntarily with those improvements – so that this year we are approaching 40 percent of the crude tank cars in use by our industry that exceed the current federal safety standard. With this achievement please keep in mind, as Administrator Quarterman said at last week's House hearing on rail safety, "getting a new tank car is not a silver bullet; first we need to prevent derailments."

In the final analysis, the women and men of the rail and oil and natural gas industries, as well as the communities traversed by our nation's freight rails, deserve our laser focus on this challenge. Our potential as a global energy leader is rooted in our ability to safely transport our game-changing energy resources safely every time, be it by truck, barge, pipeline, or rail.

We look forward to continuing to work with our colleagues in government, in the rail industry, and with others to develop a long-term, holistic and comprehensive approach to rail safety that includes prevention, mitigation, and response and moves us closer to what I think is our shared goal of zero rail incidents.

Thank you for your time and attention.

In: Transporting Crude Oil by U.S. Rail
Editor: Rosario S. McLaughlin ISBN: 978-1-63117-837-5

Chapter 8

EMERGENCY RESTRICTION/PROHIBITION ORDER, DOCKET NO. DOT-OST-2014-0025 [STANDARDS TO TRANSPORT CRUDE OIL BY RAIL]*

Anthony R. Foxx

This notice constitutes an Emergency Restriction/Prohibition Order (Order) by the United States Department of Transportation (DOT) pursuant to 49 U.S.C. § 5121(d). This Order is issued to all persons who offer for transportation in commerce within the United States, UN 1267, Petroleum crude oil, 3, Packing Group (PG) I, II, or III, as described by 49 CFR § 172.101 of the Hazardous Materials Regulations (HMR); 49 CFR Parts 171 to 180). By this Order, DOT is:

- mandating the proper testing (conducted with sufficient frequency and quality) and classification of petroleum products (i.e., petroleum crude oil) prior to them being offered into transportation;
- requiring persons who offer bulk quantities of petroleum crude oil for transportation in commerce by rail to treat Class 3 petroleum crude oil as a Packing Group (PG) I or Packing Group II hazardous material only until further notice.

* This is an edited, reformatted and augmented version of an Emergency Order issued by the U.S. Department of Transportation on February 25, 2014.

Upon information derived from recent railroad accidents and subsequent investigations and testing, the Secretary of Transportation has found that violations of the Federal Hazmat law (51 U.S.C. §§ 5101, *et seq.*) or the Hazardous Materials Regulations (HMR) (49 CFR Parts 171 to 180), and unsafe practices related to the classification and packaging of Petroleum crude oil, are causing or otherwise constitute an imminent hazard to the safe transportation of UN 1267, Petroleum crude oil, 3, PG I, II, or III. For more detailed information, see "Background/Basis for Order" below.

EFFECTIVE IMMEDIATELY ANY PERSON IDENTIFIED BY THIS ORDER

1) Shall, prior to offering into transportation, ensure that the petroleum product (i.e., petroleum crude oil) is properly tested and classed under current regulations, in accordance with the requirements of 49 CFR parts 172 and 173.
2) Shall transport by rail UN 1267, Petroleum crude oil, 3, PG I, II, or III, according to the requirements for petroleum crude oil, classified as Packing Group I or Packing Group II.

This Order applies to all persons who offer for transportation or transport by rail in commerce within the United States, petroleum crude oil, and its officers, directors, employees, subcontractors, and agents.

This Order is effective immediately and remains in effect unless withdrawn in writing by the Secretary, or until it otherwise expires by operation of regulation and/or law.

Jurisdiction

The Secretary of Transportation has the authority to regulate the transportation of crude oil in commerce. 49 U.S.C. § 5103(b). The Secretary of Transportation has designated petroleum crude oil, UN 1267, 3, Packing Group I, II, or III, as a hazardous material subject to the requirements of the HMR. 49 U.S.C. § 5121(d); 49 U.S.C. § 5103(a). Persons who offer for transportation, or transport, petroleum products (i.e., petroleum crude oil) in commerce within the United States are "persons," as defined by 49 U.S.C. §

5102(9), in addition to being "persons" under 1 U.S.C. § 1 and 49 CFR § 171.8. Commerce is as defined by 49 U.S.C. § 5102(1) and 49 CFR § 171.8, and "transportation" or "transport" are as defined by 49 U.S.C. § 5102(13) and 49 CFR 171.8. A "railroad" is as defined by 49 CFR § 171.8. A "train" is as defined by 49 CFR § 171.8. Accordingly, persons who offer for transportation or transport petroleum crude oil in commerce, including by rail, are subject to the authority and jurisdiction of the Secretary, including the authority to impose emergency restrictions, prohibitions, recalls, or out-of-service orders, without notice or an opportunity for hearing, to the extent necessary to abate the imminent hazard. 49 U.S.C. § 5121(d).

Background/Basis for Order

An imminent hazard, as defined by 49 U.S.C. 5102(5), constitutes the existence of a condition relating to hazardous material that presents a substantial likelihood that death, serious illness, severe personal injury, or a substantial endangerment to health, property, or the environment may occur before the reasonably foreseeable completion date of a formal proceeding begun to lessen the risk that death, illness, injury or endangerment.

Misclassification is one of the most dangerous mistakes to be made when dealing with hazardous materials because proper classification is the critical first step in determining how to pack, handle, communicate about, and safely transport hazardous materials. Misclassification may indicate larger problems with company management, oversight, and quality control. Petroleum crude oil may contain dissolved gases, may exhibit corrosive properties and also may exhibit toxic properties. Additionally, the flammability of crude oil being shipped by bulk rail poses a significant risk of substantial endangerment to health, property, or the environment when an explosion occurs.

In light of continued dangers associated with petroleum crude oil shipments by rail, the actions described in this Order are necessary to eliminate unsafe conditions and practices that create an imminent hazard to public health and safety and the environment.

A. Recent Crude Oil Incidents

The United States has experienced a dramatic growth in the quantity of petroleum crude oil being shipped by rail in recent years. The growth has largely been spurred by developments in North Dakota, where the Bakken formation in the Williston Basin has become a major source for oil production in the United States. Much of the Bakken petroleum crude oil is shipped via rail to refineries located near the U.S. Gulf Coast or to pipeline connections, primarily located in Oklahoma.[1] Shipping hazardous materials is inherently dangerous. Transporting petroleum crude oil can be problematic if released into the environment because it both is flammable and causes oil spills. This risk of flammability is compounded in the context of rail transportation because petroleum crude oil is commonly shipped in large unit trains. With the rising demand for rail carriage of hazardous materials, occasioned by the current demand for petroleum crude oil[2] throughout the United States, the risk of rail incidents increases. Several accidents since last summer, which caused deaths, injuries and significant property damage, have demonstrated the need for emergency action to address unsafe practices in the shipment of petroleum crude oil by rail. The shipments are an imminent hazard when Packing Group I or Packing Group II petroleum crude oil is misclassified as Packing Group III, which could lead to an improper package being used to transport the misclassified hazardous material.

Most recently, on December 30, 2013, a westbound grain train derailed 13 cars near Casselton, North Dakota,[3] fouling main track 2. Simultaneously, an eastbound petroleum crude oil unit train was operating on main track 2. The petroleum crude oil unit train reduced its speed and collided with the derailed car that was fouling main track 2, resulting in the derailment of the head-end locomotives and the first 21 cars of the petroleum crude oil unit train. Eighteen of the 21 derailed tank cars ruptured, and an estimated 400,000 gallons of crude was released. The ruptured tank cars ignited causing an explosion. Approximately 1,400 people were evacuated. Damages from the derailment have been estimated at $8 million.[4]

On November 8, 2013, a 90-car petroleum crude oil train derailed in a rural area near Aliceville, Alabama. The petroleum crude oil shipment had originated in North Dakota, and was bound for Walnut Hill, Florida, to be transported by a regional pipeline to a refinery in Saraland, Alabama. More than 20 cars derailed and the petroleum crude oil of at least 11 cars ignited resulting in an explosion and fire. Although there were no reported injuries, an undetermined amount of petroleum crude oil escaped from derailed cars fouling a wetlands area near the derailment site, and the costs are estimated at $3.9 million.

On July 6, 2013, a catastrophic railroad accident involving a U.S. railroad company occurred in Lac-Mégantic, Quebec, Canada, when an unattended freight train transporting crude oil rolled down a descending grade and subsequently derailed.[5] The derailment resulted in multiple explosions and subsequent fires, which caused the confirmed death of forty-two people and presumed death of five more, extensive damage to the town center and the evacuation of approximately 2,000 people from the surrounding area. Preliminary estimates of costs exceed $1 billion.

B. DOT Actions and Investigation of Petroleum Crude Oil Classification

In the wake of these and other events, Pipeline and Hazardous Materials Safety Administration (PHMSA) and Federal Rail Administration (FRA) have taken a number of steps to increase the safety of petroleum crude oil shipments by rail. Following the Lac-Mégantic derailment, on August 7, 2013, FRA issued Emergency Order No. 28 (EO 28), establishing securement requirements for certain unattended trains and rail equipment, including petroleum crude oil unit trains. EO 28 remains in effect until further notice by FRA. In addition, on August 7, 2013, PHMSA and FRA issued Safety Advisory 2013-06, which made a number of safety-related recommendations to railroads and hazardous materials offerors operating in the United States, including the recommendation that offerors evaluate their processes to ensure that hazardous materials are properly classed and described in accordance with the HMR, and the recommendation that offerors and carriers conduct reviews of their safety and security plans. On August 27-28, 2013, FRA and PHMSA held a public meeting with industry stakeholders to solicit input on a comprehensive review of safety regulations contained in 49 CFR Part 174 applicable to the safe transportation of hazardous materials by rail. PHMSA and FRA are collaborating to address comments received at the public meeting.

On August 29, 2013, FRA convened an emergency session of the Railroad Safety Advisory Committee (RSAC). RSAC is a group composed of railroad industry, labor, and governmental representatives who develop recommendations on new regulatory standards and other rail safety programs. During the emergency meeting, RSAC established three collaborative working groups to formulate new rulemaking recommendations regarding hazardous materials transportation by rail, appropriate train crew sizes, and train securement procedures. Each of these working groups has been meeting on a regular basis and each working group is expected to produce formal recommendations for consideration on or before April 2014.

On September 6, 2013, PHMSA issued an Advanced Notice of Proposed Rulemaking (ANPRM (HM-251); 78 Fed. Reg. 54849) to solicit comments on petitions for rulemaking and National Transportation Safety Board (NTSB) recommendations related to rail hazmat safety, including regulations regarding operational practices and DOT specification tank cars, most commonly used to move crude oil by rail. The comment period closed on December 5, 2013, and PHMSA received 135 comments representing almost 100,000 stakeholders. The Department is actively working on developing a regulatory proposal.

PHMSA and FRA issued a supplementary safety advisory, Safety Advisory 2013-07, on November 20, 2013, to emphasize the importance of proper characterization, classification, and selection of a packing group for Class 3 materials (flammable liquids, including petroleum crude oil), and to reinforce the need to follow the Federal hazardous materials regulations for safety and security planning.

On January 16, 2014, the Secretary of Transportation met with members of the rail and the petroleum industries in a call to action to address the risks associated with the transportation of crude oil by rail.

Notwithstanding the above DOT actions, in light of continued risks associated with petroleum crude oil shipments by rail, the further action described in this Order is necessary to eliminate unsafe conditions and practices related to the classification and packaging of petroleum crude oil that create an imminent hazard to public health and safety and the environment.

C. Classification and Packaging of Petroleum Crude Oil

The proper classification and characterization of a hazardous material is critical and required under the HMR by Parts 171, 172 and 173, as it dictates additional requirements, such as operational controls, emergency response, and proper packaging selection. The HMR is essential for safe transportation. The classification requirements in the HMR dictate the appropriate and authorized selection of packaging, fill densities and outage; accompanying hazard communications (markings, labels and placards); transportation safety and operational controls; and safety and security planning. Not only can properly classified shipments mitigate injury and damage when accidents occur, they are necessary to enable the most effective and informed emergency response. Indeed, the accurate classification of hazardous materials in transportation can protect the safety of emergency responders and others who may come into contact with the material.

Proper classification and characterization is especially important when dealing with an organic material such as mined liquids and gases, as these materials have variable characteristics, unlike manufactured products which are generally consistent. Moreover, crude oil transported by rail often derives from different sources and is then blended, further enhancing the potential range of appropriate classifications.

It is the offeror's responsibility to properly classify and describe the hazardous material in accordance with parts 172 and 173 of the HMR. *See* § 171.1 and 171.2 and 173.22. When a single material meets more than one hazard class, the shipper must select the proper shipping name based on the hazard precedence table in § 173.2a. Once an offeror has determined the hazard class of the material, the offeror must select the most appropriate proper shipping name from the HMR.

In the case of petroleum crude oil, minimum properties to properly test, characterize and classify the material include: flash point, corrosivity, specific gravity at loading and reference temperatures, and the presence and concentration of specific compounds such as sulfur. This information enables an offeror to properly classify a hazardous material and select the proper packaging from the list of packaging authorized for use by the HMR. Such information and determination of the authorized packaging also ensure that the appropriate outage is maintained in accordance with § 173.24b.

With regard to package selection, the HMR requires at § 173.24(b) that each package used for the shipment of hazardous materials shall be designed, constructed, maintained, filled, its contents so limited, and closed, so that under conditions normally incident to transportation there will be no identifiable (without the use of instruments) release of hazardous materials to the environment, and further requires that the effectiveness of the package will not be substantially reduced. Under this requirement, offerors must consider how the properties of the material, including temperature and pressure, may affect the packaging.

PHMSA and FRA issued a supplementary safety advisory, Safety Advisory 2013-07, on November 20, 2013, to emphasize the importance of proper characterization, classification, and selection of a packing group for Class 3 materials, and to reinforce the need to follow the Federal hazardous materials regulations for safety and security planning.

In addition, PHMSA and FRA initiated Operation Classification in August 2013, which involved unannounced inspections requesting samples of the transported petroleum crude oil and testing the oil samples to verify that offerors of the materials have properly classified and described the hazardous materials. Preliminary testing has focused on the classification and packing group assignments that had been selected. These tests measured some of the inherent chemical properties of the petroleum crude oil collected. PHMSA tested samples of petroleum crude oil, destined for transportation by rail, and determined that some of the samples had been erroneously assigned to a lower packing group by offerors than required under the HMR. As a result of

Operation Classification, PHMSA has proposed $93,000 in civil penalties against petroleum crude oil offerors. PHMSA and FRA will continue to test the Bakken petroleum crude oil characteristics to gain information to verify whether the crude oil appears to be classified appropriately under the hazardous materials regulations.

FRA has also discovered incidents, and brought enforcement action, where petroleum crude oil was misclassified. FRA audits of crude oil loading facilities indicate that the classification of crude oil being transported by rail is often based solely on Materials Safety Data Sheet (MSDS) data that provide only a material classification and a range of material properties. This MSDS information is typically provided by the consignee to the shipper, and the shipper does not validate the values of the crude oil properties. FRA's audits indicate that MSDS information has been based upon old test data, and not from testing for the many different sources of the crude oil. At a petroleum crude oil rail trans-loading facility, the crude oil is collected and stored in an above-ground intermediate storage tank. The above ground intermediate storage tanks accumulate hundreds of cargo tank motor vehicle loads, which may also lead to classification issues. In a problematic example, a shipper provided information to FRA showing that crude oil being transported by rail had a flash point of 68°F, or a Packing Group I hazardous material. The crude oil had been improperly classified as a Packing Group III material, and was being transporting in AAR Specification tank cars that were not equipped with the required design enhancements for DOT Specification 111 cars. This constituted a misuse of the crude oil HMR packaging exceptions, and subsequent violations of the HMR. Reliance upon MSDS information that is outdated and that does not accurately represent the hazardous material may also lead to the incorrect packing group being assigned to the hazardous material, which can lead to an unauthorized package being used.

Misclassification can also lead to use of unauthorized containers that lack the required safety enhancements necessary to safely transport PG I & II materials, as well as insufficient development of safety and security plans and the communication of inaccurate information to emergency responders. As a result, these conditions pose an imminent hazard.

D. Investigations Conducted by Other Agencies Involving Petroleum Crude Oil

Several Government entities have conducted research on transporting petroleum crude oil on trains. These entities include: the United States Department of State, Transport Canada, Transportation Safety Board of

Canada, Congressional Research Service, and the National Transportation Safety Board.

While conducting its review of the National Interest Determination for the Keystone Pipeline, on January 31, 2014, the State Department released its Final Supplemental Environmental Impact Statement. This study reviewed the alternatives to building the Keystone Pipeline, which would service the Bakken formation of the Williston Basin. In determining the no-action alternative scenario, in which a pipeline is not built and unit trains continue to deliver the petroleum crude oil, the State Department determined that "a projection of injury and fatality frequencies onto the crude oil transport volume … indicates a potential additional 49 injuries and six fatalities for the rail alternative"[6] per year.

The Congressional Research Service (CRS), a nonpartisan research group for the United States Congress,[7] has analyzed the U.S. Rail Transportation of Crude Oil issue and on February 6, 2014, prepared a report for Congress.[8] In its Report, CRS determined that the increasing deployment of unit trains concentrates a large amount of potentially environmentally harmful and flammable material, increasing the probability that, should an accident occur, large fires and explosions could result.[9] Further, the increased use of rail for crude oil shipments is likely to increase the number of incidents, some of which may involve oil spills.[10]

Transport Canada (TC), the Canadian equivalent to the U.S. Department of Transportation, has implemented many changes as a result of the catastrophic accident in Lac-Mégantic on July 6, 2013. Most recently, TC has implemented Protective Direction No. 31.[11] This Directive requires that any person engaged in importing or offering crude oil for transport to immediately test the classification of crude oil being imported, handled, offered for transport or transported as UN 1267, or UN 1993, if the classification testing has not been conducted since July 7, 2013, and to provide those test results to TC upon request. Moreover, following testing, any person who imports or offers for transport UN 1267 or UN 1993 must immediately provide a MSDS for the tested product to TC, through the Canadian Transport Emergency Centre, also known as CANUTEC. Finally, until such time as the person importing or offering for transport has completed the testing, any person who imports, handles, offers for transport, or transports crude oil classified as UN 1267 or UN 1993 by rail must ship all such crude oil as a Class 3 Flammable Liquid PG I and meet the requirements established in the Act, regulations and standards regarding UN 1267 or UN 1993 classified as PG I. Classification testing must follow the requirements of the Act and its regulations.

Finally, the National Transportation Safety Board has made recommendations to PHMSA and FRA on January 23, 2014.[12] In its recommendations to PHMSA regarding the transportation of petroleum crude oil, the NTSB indicated Canadian authorities analyzed petroleum crude oil from nine undamaged tank cars in the train that caused the Lac-Mégantic accident.[13] Testing indicated that the crude oil being transported in those trains had been incorrectly assigned to Packing Group III rather than Packing Group II.[14] The crude oil on the Lac-Mégantic accident train was derived from wells in the Bakken.[15] The above information is clear evidence of an ongoing problem with classification of petroleum crude oil that is being shipped by rail. The continuing misclassification has consequences with regard to the type of packaging in which the petroleum crude oil is transported, the protections those packages provide if an accident occurs, and, notably with regard to emergency responders, sufficient knowledge about the hazards of the materials being transported so that if an accident occurs, they can respond appropriately. The Secretary finds that this misclassification of petroleum crude oil as a Packing Group III is an imminent hazard (as defined by § 109.1) that presents a substantial likelihood that death, serious illness, severe personal injury, or a substantial endangerment to health, property, or the environment may occur. The information conveyed to PHMSA by the NTSB in its recent recommendations regarding the misclassification of the petroleum crude oil in the Lac-Mégantic accident train provides evidence of this imminent hazard.

NTSB also issued recommendation R-14-6 that PHMSA "require shippers [offerors] to sufficiently test and document the physical and chemical characteristics of hazardous materials to ensure the proper classification, packaging, and record-keeping of products offered in transportation."[16] In light of the above discussion, PHMSA concurs with NTSB, and to address the imminent hazard, this Order requires that any person who offers a large bulk quantity of petroleum crude oil into transportation in commerce conduct minimum testing of that petroleum crude oil to verify its classification and to retain records of that testing. Each offeror must conduct testing of a sample of a bulk quantity of petroleum crude oil being offered into transportation in commerce prior to offering that bulk quantity of petroleum crude oil into transportation. At a minimum, the tests shall be capable of determining the petroleum crude oil's flash point; boiling point; corrosivity to steel and aluminum; presence and content of compounds such as sulfur/hydrogen sulfide; percentage presence of flammable gases; and, the vapor pressure at 50°C. The offeror should clearly identify the method used to determine the vapor pressure.

This testing will also help ensure that petroleum crude oil offerors are testing the hazardous material regularly and verifying that they have been classified, described and packaged the petroleum crude oil in accordance with the HMR. This minimum testing will address, in part, the imminent hazard described above. Offerors' review of the testing data could help DOT to ensure compliance with existing HMR requirements regarding the proper classification of petroleum crude oil and the subsequent selection of permissible packaging.

As discussed in PHMSA's January 2, 2014 safety alert, PHMSA is still investigating the variability and flammability of petroleum crude oil being produced in the Bakken. Because the evidence indicates that petroleum crude oil being produced in the Bakken is being misclassified and improperly packaged in non-DOT Specification tank cars, and because the properties and volatility of this petroleum crude oil are still unknown in certain regards, the Secretary finds that the Order is necessary to prevent the imminent hazard that exists when petroleum crude oil being transported by railroad is misclassified. This Order ensures that petroleum crude oil will be transported (until this Order is rescinded, amended or otherwise expires by regulation or law) in, at a minimum, a DOT Specification tank car.

Remedial Action

To eliminate or abate the imminent hazard, you must:

1. prior to offering into transportation, or transporting, ensure that the petroleum crude oil is properly tested and classed in accordance with 49 CFR parts 172 and 173.
2. From the date of this Order, assure that shipments by rail of UN 1267, petroleum crude oil, 3, PG I, II, or III shall be handled as a Packing Group I or II hazardous material.

Rescission of this Order

Before you may offer and/or transport any package subject to this order you must be able to adequately demonstrate that you have taken the actions listed above, or that you have taken other actions, and that the actions taken have, in fact, resulted in an imminent hazard no longer existing, or if a change in applicable statute or Federal regulation occurs that supersedes the requirements of this Order, the Secretary will issue a Rescission Order.

Failure to Comply

Any person failing to comply with this Emergency Order is subject to civil penalties of up to $175,000 for each violation or for each day they are found to be in violation (49 U.S.C. §5123). A person violating this Emergency Order is also subject to criminal prosecution, which may result in fines under title 18, imprisonment of up to ten years, or both (49 U.S.C. § 5124).

Right to Review

Pursuant to 49 U.S.C. § 5121(d)(3) and in accordance with section 554 of the Administrative Procedure Act (APA), 5 U.S.C. §§ 500 *et seq*, a review of this action may be filed. Any petition seeking relief must be filed within 20 calendar days of the date of this order (49 U.S.C. § 5121 (d)(3)), and addressed to U.S. DOT Dockets, U.S. Department of Transportation, 1200 New Jersey Avenue, S.E., Room W12-140, Washington, DC 20590 (http://Regulations.gov). Furthermore, a petition for review must state the material facts at issue which the petitioner believes dispute the existence of an imminent hazard and must include all evidence and exhibits to be considered. The petition must also state the relief sought. Within 30 days from the date the petition for review is filed, the Secretary must approve or deny the relief in writing; or find that the imminent hazard continues to exist, and extend the original Emergency Order. In response to a petition for review, the Secretary may grant the requested relief in whole or in part; or may order other relief as justice may require (including the immediate assignment of the case to the Office of Hearings for a formal hearing on the record).

End Notes

[1] See AAR's December 2013 paper "Moving Petroleum crude oil by Rail", available online at: https://www.aar.org/keyissues/Documents/Background-Papers/Crude-oil.

[2] In 2011 there were 65,751 originations of tank car loads of crude oil. In 2012, there were 233,811 originations. Association of American Railroads, *Moving Crude Petroleum by*

Rail, https://www.aar.org/keyissues/Documents/Background-Papers/Moving%20Crude%20Petroleum%20by%20Rail%202012-12-10.pdf (December 2012).

[3] This derailment currently is being investigated by the National Transportation Safety Board (NTSB), and information regarding this incident can be found at the NTSB website. See http://www.ntsb.gov/doclib/reports/2014/Casselton_ND_Preliminary.pdf.

[4] *Federal Rail Administration*

[5] This derailment currently is being investigated by the Transportation Safety Board of Canada and information regarding this incident can be found at the TSB website. See http://www.bst-tsb.gc.ca/eng/enquetes-investigations/rail/2013/R13D0054/R13D0054.asp

[6] Found at: http://keystonepipeline-xl.state. While the State Department's analysis did not take into account any measures that FRA and PHMSA have taken and are taking with respect to the safety of rail transport of crude oil (which would tend to increase projected injuries and fatalities), they also did not consider the Lac-Mégantic accident, which was outside of the time frame that the State Department considered (and would tend to decrease projected injuries and fatalities). Chapter 5.1.1.4 p. 5.1-6

[7] The Congressional Research Service (CRS) works exclusively for the United States Congress, providing policy and legal analysis to committees and Members of both the House and Senate, regardless of party affiliation. As a legislative branch agency within the Library of Congress, CRS has been a valued and respected resource on Capitol Hill for nearly a century.

[8] http://www.fas.org/sgp/crs/misc/R43390.pdf

[9] *Id.*, at p. 10.

[10] *Id.*, at p. 19.

[11] Found at: http://www.tc.gc.ca/eng/mediaroom/backgrounders-protective-direction-no31-7385.html

[12] Found at: http://www.ntsb.gov/doclib/recletters/2014/R-14-001-003.pdf and http://www.ntsb.gov/doclib/recletters/2014/R-14-004-006.pdf.

[13] http://www.ntsb.gov/doclib/recletters/2014/R-14-004-006.pdf. The Transportation Safety Board of Canada issued a press release relating to the misclassification on September 11, 2013, http://www.bst-tsb.gc.ca/eng/medias-media/communiques/rail/2013/r13d0054-20130911.asp.

[14] *Id.*

[15] *Id.*

[16] *Id.*

INDEX

#

A

B

C

D

E

F

G

H

I

J

K

L

M

N

O

P

T

U

V

W

Y